Television Performance

Television
Performance

Edited by Lucy Fife Donaldson and James Walters

First published 2019 by
RED GLOBE PRESS

Red Globe Press in the UK is an imprint of Springer Nature Limited, registered in England, company number 785998, of 4 Crinan Street, London N1 9XW.

Red Globe Press® is a registered trademark in the United States, the United Kingdom, Europe and other countries.

ISBN 978–1–137–60820–8 hardback
ISBN 978–1–137–60819–2 paperback

This book is printed on paper suitable for recycling and made from fully managed and sustained forest sources. Logging, pulping and manufacturing processes are expected to conform to the environmental regulations of the country of origin.

A catalogue record for this book is available from the British Library.

A catalog record for this book is available from the Library of Congress.

The editors and publisher would also like to thank Glen Creeber for permission to use copyright text material: the extract in Chapter 3 from *Serial Television: Big Drama on the Small Screen* (British Film Institute Publishing, 2004).

Contents

List of Figures

Acknowledgements

We would like to thank our contributors, Jonathan Bignell, Tom Brown, Lydia Buckingham, Sarah Cardwell, Gary Cassidy, Alex Clayton, Amy Holdsworth, Simone Knox, Elliott Logan, Karen Lury, Timotheus Vermeulen and James Zborowski, for their dedication to this collection's aims, and for providing us with such a rich array of essays. We would also like to thank the team at Red Globe Press (formerly Palgrave HE) for their care and consideration throughout the publication process, and Cathy Tingle for her meticulous copy-editing work. Finally, we would like to recognise the great debt we owe to Andrew Klevan, with whom we both studied performance for the first time as undergraduates. Without Andrew's profound influence, it is unlikely that we would have sought to produce this book on television performance and it is debatable whether we would have pursued academic careers at all. We remain forever grateful to Andrew for sharing his ideas with us and for supporting us in our pursuits. This book is dedicated to him.

Lucy Fife Donaldson
James Walters

Contributors

Jonathan Bignell is Professor of Television and Film at the University of Reading. His books include *Writing and Cinema* (Taylor & Francis, 1999); three editions of *An Introduction to Television Studies* (Routledge, 2004, 2007, 2012); *Beckett on Screen* (Manchester University Press, 2009); *A European Television History* (edited with Andreas Fickers) (Wiley-Blackwell, 2008); *Big Brother: Reality TV in the Twenty-first Century* (Palgrave Macmillan, 2005); and *Postmodern Media Culture* (Edinburgh University Press, 2000). His articles about television include contributions to the journals *Critical Studies in Television*; the *Historical Journal of Film, Radio and Television*; *Media History*; and *Screen*. His recent work has been on science fiction TV of the 1960s, and the history of transatlantic television drama.

Tom Brown is Senior Lecturer in Film Studies at King's College London. He is the author of *Spectacle in "Classical" Cinemas: Musicality and Historicity in the 1930s* (Routledge, 2015) and *Breaking the Fourth Wall: Direct Address in the Cinema* (Edinburgh University Press, 2012). He is the co-editor of three books: *The Biopic in Contemporary Film Culture* (Routledge, 2014), *Film Moments: Criticism, History, Theory* (BFI/Palgrave, 2010) and *Film and Television after DVD* (Routledge, 2008).

Lydia Buckingham teaches stardom and performance at King's College London, where she is in the final year of her PhD. She is the author of the forthcoming book chapter 'The Body and the Woman: Contemporary Hollywood, Actresses and Nudity' (Cambridge Scholars Press).

Sarah Cardwell is Honorary Fellow in the School of Arts, University of Kent. She is the author of *Adaptation Revisited* (Manchester University Press, 2002) and *Andrew Davies* (Manchester University Press, 2005), as well as numerous articles and papers on film and television aesthetics, literary adaptation, contemporary British literature, and British cinema and television. She is a founding co-editor of 'The Television Series' (Manchester University Press), Editorial Advisor for *Critical Studies in Television*, and on the advisory board for the new series 'Adaptation and Visual Culture' (Palgrave Macmillan).

Gary Cassidy is a Senior Lecturer in Acting at Bath Spa University. He has recently completed his AHRC-funded doctoral research at the University of Reading. His thesis explored the rehearsal process of playwright Anthony Neilson, using filmed footage of rehearsals and interviews. He trained as an actor at the Royal Scottish Academy of Music and Drama (now the Royal Conservatoire of Scotland) and has many years of professional acting experience (under his Equity name, Cas Harkins) covering film, theatre, television and radio. He has published in the *International Journal of Scottish Theatre and Screen*, and co-authors the blog strand 'What Actors Do' for *CST Online*.

Alex Clayton is Senior Lecturer in Film and Television at the University of Bristol. He is the author of *The Body in Hollywood Slapstick* (McFarland, 2007), co-editor of *The Language and Style of Film Criticism* (Routledge, 2011), and a member of the editorial board of *Movie: A Journal of Film Criticism*. He has published a range of essays on screen comedy, performance and aesthetics, and his next book is called *Funny How? Sketch Comedy and the Art of Humor* (SUNY Press, forthcoming).

Lucy Fife Donaldson is Senior Lecturer in Film Studies at the University of St Andrews and her research focuses on the materiality of style and the body in popular film and television. She is the author of *Texture in Film* (Palgrave Macmillan, 2014), and a member of the editorial board of *Movie: A Journal of Film Criticism*.

Amy Holdsworth is Senior Lecturer in Film and Television Studies at the University of Glasgow. She is the author of *Television, Memory and Nostalgia* (Palgrave Macmillan, 2011) and has contributed articles on television theory and aesthetics to *Screen*; *Cinema Journal*; *Critical Studies in Television*; *Journal of Popular Television*; and the *Journal of British Cinema and Television*.

Simone Knox is Associate Professor of Film and Television at the University of Reading. She sits on the board of editors for *Critical Studies in Television* and her publications include essays in *Film Criticism*; *Journal of Popular Film and Television*; *New Review of Film and Television Studies*; the *Historical Journal of Film, Radio and Television*; and the *Journal of British Cinema and Television*. She co-authors, with Gary Cassidy, the blog strand 'What Actors Do' for *CST Online*.

Elliott Logan is at the University of Queensland. He is the author of *Breaking Bad and Dignity* (Palgrave Macmillan, 2016), and has published essays on film and television aesthetics in *Screen*; *Critical Studies in Television*; and *New Review of Film and Television Studies*. He is Associate Editor of the journal *Series*.

Karen Lury is Professor of Film and Television Studies at the University of Glasgow. She has published widely on film and television, specifically in relationship to representations of the child and childhood. She is the author of *British Youth Television* (Oxford University Press, 2001), and *Interpreting Television* (Bloomsbury, 2005) and is an editor of the international journal *Screen*.

Timotheus Vermeulen is Associate Professor in Media, Culture and Society at the University of Oslo, Norway. His research interests include cultural theory, aesthetics, and close textual analysis of film, television and contemporary art. Vermeulen is the author of multiple books and has edited various anthologies and special journal issues, most recently – with Catherine Constable and Matt Denny – an issue on surface aesthetics in *Film-Philosophy*. He publishes in academic and popular contexts alike, writing for, among others, *The Journal of Aesthetics and Culture*; *e-flux*; *Screen*; *Monu*; *The American Book Review*; *Texte Zur Kunst*; *Tank*; and *Metropolis M*, as well as various reference works, collections and catalogues. He is a regular contributor to *Frieze*.

James Walters is Reader in Film and Television Studies at the University of Birmingham. His books include *Alternative Worlds in Hollywood Cinema* (Intellect/Chicago University Press, 2008); *Film Moments: Criticism, History, Theory* (BFI/Palgrave, 2010); *Fantasy Film* (Bloomsbury, 2011); and the BFI Television Classic on *The Thick of It* (2016).

James Zborowski is Senior Lecturer in Film and Television Studies at the University of Hull. His previous publications include an exploration of methodological issues in television aesthetics; a critical survey of the work of Sally Wainwright; an analysis of contemporary British lifestyle television; and articles and chapters on various individual television programmes: *Ghostwatch*; *The Royle Family*; *The Simpsons*; and *The Wire*. His first book, *Classical Hollywood Cinema: Point of View and Communication*, was published in 2016 by Manchester University Press.

Critical Introduction

Lucy Fife Donaldson and James Walters

Performance is central to our understanding and appreciation of television. It is also the case, however, that the range of performance types on television is especially broad and varied, encompassing work that is specialised and amateur, stylised and natural, overt and opaque. The actor contributes to a diverse set of genres and forms of programming, from the one-off television play to the soap opera, situation comedy to science fiction, period drama to political satire. They may provide voice-over work on a natural history series or documentary exposé. They can promote themselves through guest appearances on chat shows or promote products in advertisements. Other professionals – such as comedians, singers, presenters, newsreaders – are required to draw upon a range of performance skills in order to demonstrate their vocation, to conduct interviews, deliver information or even guide an audience through a series of games. Non-professionals also contribute to the variety of television programming, whether as amateur performers or through being observed in structured or unstructured documentary and reality formats. As a result, an inexhaustible wealth of material is made available to the audience and to the critic, offering many different opportunities to engage with performance on television. The chapters in this collection propose a variety of approaches to the layers of what might be defined and evaluated as performance, responding to the wide-ranging work of individuals and groups performing on television. Each of the book's contributions seeks to value what is special and particular about television performance and, as part of that effort, develop the critical language that is building around the study of performance on television.

Starting with Performance

We can begin to frame these wider interests by turning briefly to an example that encapsulates some of the complexities involved in defining and critically evaluating television performance. The comedy sketch show

Key and Peele (Comedy Central, 2012–15), fronted by a pair of performers (Keegan-Michael Key and Jordan Peele, who also write for the show), offers a particular set of opportunities for performance that are shaped by specific qualities of television. Each episode combines several short sketches interlinked by sections where Key and Peele perform quasi stand-up to a live studio audience (seasons one to three) or talk to one other in a car as they drive through the desert (seasons four and five).[1] As a result, the time of performance is both shortened – condensed into sketches lasting only a few minutes – and extended, through the episodic structure of the programme and in certain serialised sketches that develop across the programme's run (perhaps most famously, President Obama and his 'anger translator' Luther). The temporal limitations of the series are emphasised through short bursts of access to the fictional world of a sketch (even when these might be repeated across the series), while the interstitial scenes of the duo performing as themselves offer brief punctuation between sketches. The space of performance is both continuous, held together by the primary setting of the studio/car to which we continually return, and markedly fragmented through the movement between varied settings, many of which are never repeated.

The moments between sketches offer an 'in-between' in both space and time, a place where the fictional worlds of the sketches are suspended and performance fills the gap left by an absence of narrative or consistency of fictional space. Key and Peele often comment on a facet of the sketch just shown (or to come), their framing or contextualising of jokes highlighting the construction of the sketches. The pair's gestures and movements connect to the audience and to one another, as they move between looking out and looking between themselves, posing forward and turning inwards. These shifts move them between performing *with* – both men frequently use similar hand movements, so that their gestures often find them paralleling one another (Figure 0.1) – and performing *to*. As this fluctuating pattern progresses, their responses and bodily postures intensify. In the very first of these segments, Key and Peele engage in a sequence of one-upmanship using the joke formulation 'we sound whiter than…', which enlarges the gestures out to the audience through repetition and embellishment, but also brings them closer together as they each seek to make the other laugh. The interstitial scenes also dramatise their differences, especially when performed on a stage large enough for one performer to execute a brief explosion of energy in response to a joke made by the other. Typically the eruption comes from

Figure 0.1 Keegan-Michael Key and Jordan Peele in the studio between sketches (*Key and Peele*, Comedy Central, 2012–15)

Key, who in this instance doubles over, clapping and laughing at Peele's 'we sound whiter than Mitt Romney in a snowstorm' (Figure 0.2). Such propulsions of energy feed back into the timeliness of these moments, endowing

Figure 0.2 Key doubles over in response to Peele (*Key and Peele*, Comedy Central, 2012–15)

them with a forward motion, while energetic movements around the set (as when Peele robotically sidesteps away from Key in response to, and as an abstractly physical expansion of, the joke) also develop the open but expressively limited blank space of the studio as a setting for play. Rather than operating as a way of filling out the pauses between sketches, then, the interstitials engage productively with their fundamental requirement to create comedic performance in a limited or limiting physical space, and to shepherd the general tone of the programme through the guidance of its makers.

We can dig further into *Key and Peele* to consider the range of opportunities for interaction that are generated for the performers, whose work often seeks out the potential for comedy to emerge within the limits of time and space. Perhaps the most extreme example of this can be found in their 'East/West College Bowl' sketch, which involves Key and Peele alternating a direct address to the camera as they embody increasingly bizarrely named college football players. The performers are presented in a tightly limited space, each in a medium shot in a studio set against colourful background graphics, and time, with each character afforded roughly five seconds for Key or Peele to distinguish them through vocal modulations and facial expressions, communicating a gamut of contrasts between laid-back laconicism, intense concentration and aggressively direct delivery (Figure 0.3). The series and

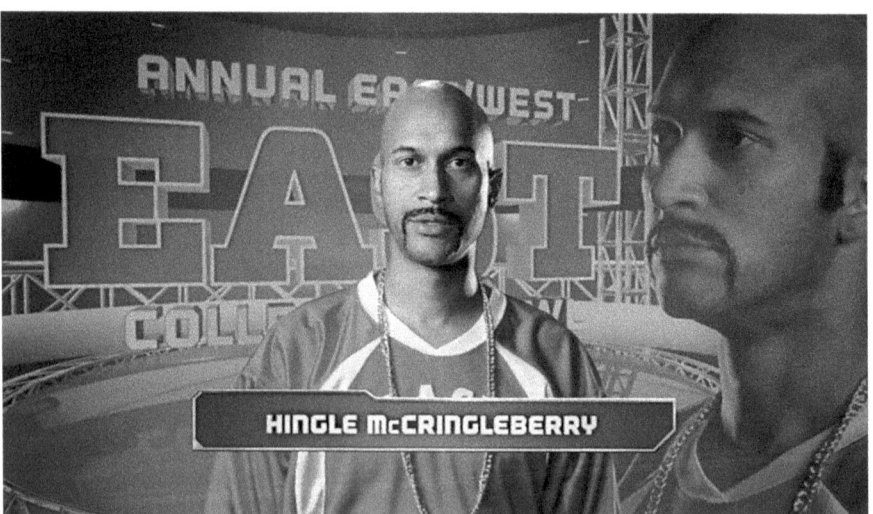

Figure 0.3 One of Key's football characters in the 'East/West College Bowl' sketch (*Key and Peele*, Comedy Central, 2012–15)

its sketches are most frequently built around comedy performances that negotiate everyday interactions and the codes of behaviour that these entail: a substitute teacher encountering middle-class white students for the first time, a town hall meeting, the delights of a continental breakfast. The performances of Key and Peele become crucial for grounding the fictional world with all its idiosyncrasies while also gesturing outwards to fill in narrative and characterisation beyond the specific moment of the sketch.

Key and Peele's first sketch in episode one provides an exemplar of the show's investment in exploring the ordinary, making an everyday occurrence comedic. The action consists of two men on phones passing one another in the street. Key waits at a corner, speaking to his wife about the theatre tickets he is planning to buy for her birthday. Peele exits a shop and calls someone to let them know he is nearby. Before Peele emerges, Key paces gently, gesturing lightly with his left hand and then holding his index finger to his ear, his features creased in a relaxed smile. However, as soon as he becomes aware of Peele, as the other man approaches to pass him, Key's eyes and mouth straighten, his voice dropping mid-sentence to a lower pitch while incorporating a vernacular drawl. He places his hand palm down on chest, fingers splayed, in a more emphatic gesture, while shifting from side to side, his body more tautly posed than when pacing moments earlier. Peele's dialogue mimics the inflection of Key's voice as he answers his phone in an even deeper tone: 'What's up dog? Imma bout five minutes away.' Peele then stands – ostensibly waiting to cross the road – and continues to speak into his phone in increasingly aggravated and brief bursts. After a few beats of the men standing next to one another, Peele turns back slightly to catch Key's gaze and both nod briefly. The scene ends with Peele moving to walk across the street, the camera moving backwards with him, while his voice shifts up to a higher and more nasal pitch commonly associated with valley-girl speak: 'Oh my god Christian, I almost totally just got mugged right nowww.'

For Key and Peele, the most mundane and incidental of spaces – a street corner – becomes an opportunity to showcase their abilities (and survival strategy as biracial men in America) to switch between coded behaviours of gender and race in a way that draws attention to the performativity of both. There is also an appreciation in the sketch of the tensions between the public and the private in these mundane spaces – how private conversations might be overheard and received in public by strangers, and how postures and tones of voice might respond to and signal those tensions, moving from relaxed to defensive. The rapidity of their contact means that the switching of behaviour must be both smooth and magnified, as with Key's deepening of his voice mid-sentence. The adjustment to his voice emphasises smoothness as words become elongated when the syllables are more forcefully pronounced – theatre becomes 'the-A-ter' – whereas his previous vocal pattern contains several stuttering hesitations. The time pressure embedded within their brief overlapping in the transitional space, where there is a narrow window available for each performer to project and move between identities before they go their separate ways, intensifies the need for fluidity and efficiency in the performances. In a mark of the subtlety contained within the amplification of gesture, accent and posture, neither performer unbalances the other, their shifts in volume, pitch and action matched. This mutuality is well expressed in the blocking that places Key slightly behind Peele, positioning the performers in a comparative tableau, so that we can see the paralleling of types of movement – shifting in pace, a scrunching of brow and mouth to project a piece of dialogue – as they continue to speak (Figure 0.4). In this short moment, the performances and their presentation are efficiently and expressively balanced so that we can appreciate the interaction between Key and Peele, their rapport being a key factor of the series as a whole and

Figure 0.4   Key and Peele pass each other on the street in their first sketch (*Key and Peele*, Comedy Central, 2012–15)

as driven more forcefully by the moments between sketches. A precise integration of performer and camera, performer and location, and between the performers themselves is achieved in order to maximise the comedic potential in a fleeting moment of exchange.

Even from this very preliminary investigation into one kind of television programme, the fundamental contribution and significance of the detailed work of the performers becomes clear. *Key and Peele* is both exemplary and typical in being a television programme led by performers whose skills deftly shape the series' structure; in this case the potentially disruptive shifts between individual sketches and between sketch and intermedial material are managed through carefully controlled adjustments in tone and energy, which bring new depth and complexity to even the briefest of interactions. In short, the work of the performers is crucial to the success of the programme. And yet, more generally, performance has not enjoyed a central position in television studies. While in the face of examples like *Key and Peele* (and those offered in this collection) this may strike us as a conspicuous oversight, it is not hard to identify similar or equivalent areas of relative neglect in the field. We might note, for example, that Manchester University Press's long-running 'Television Series' of books, edited by Sarah Cardwell, Jonathan Bignell and Steven Peacock, exists in part as an attempt to devote overdue attention towards the work of writers, while the contribution of television professionals such as directors, producers, costume designers, composers etc. can often receive only marginal consideration in critical accounts of programmes. This is perhaps indicative of a wider and continuing tendency, identified by John Corner, whereby 'direct attention to production processes has sometimes been seen as a neglected or underdeveloped aspect of enquiry' (1999: 70). Many areas of television – and the job roles involved in processes of production in particular – remain underexplored, some to the point of being barely visible in scholarly debate. It is especially curious that performance should be overlooked, however, given the strong and sustained awareness that the work of actors receives in the public appreciation of television. Actors are the creative element most likely to be featured in the promotion of programmes and, equally, they very often remain the most identifiable features for television audiences.[2] The performer is a key controlling force in shaping character (along with the writer and director, casting directors and producers, and other creative production personnel such as costume designers), and, as Jason Mittell points out, 'one of the primary ways that viewers engage with programming is to develop long-term relationships with char-acters' (2015: 127), thereby making the work of the actor in shaping their character integral to audience enjoyment and pleasure in (or even displeas-ure with) a particular programme. Moreover, an evaluation of a programme's broader achievements can often rely significantly upon the accomplishment of its performers. This recognition extends to industry awards ceremonies, where categories celebrating the work of television actors are well publicised and long established – since the 1950s for the Emmy Awards and the British Academy Television Awards, and since the 1960s for the Golden Globes. It might reasonably be concluded, therefore, that television studies has been relatively slow to pay proper consideration to something that is highly visible and widely recognised outside academic debate: the importance of television performers.

Acting/Performance

Our attention to moments from *Key and Peele* goes some way to initiat-ing this collection's shared investment in the detail of performance, in finding the right words to carefully describe and evaluate the choices made by performers on television and how these operate within the context of a particular programme, its audiovisual style. Attending to the shifting modes of performance offered by Keegan-Michael Key and Jordan Peele starts to illustrate some of the ways in which televisual performance is required to adapt to the varied spaces and modes of television, especially the ways in which performances on television shift between layers of rehearsed, pro-fessional acting technique and seemingly everyday, instinctive behaviours, working to blur distinctions between the two (whether that blurring is intentional or not). In this respect, the example of *Key and Peele* is a useful starting point for exploring the challenge of approaching a medium which incorporates and sustains multiple modes of performing.

This collection therefore aims not only to address a more general gap in the field but also to expand the approach to the question of what actors do in/for television, and in so doing take as our focus the means by which we can best appreciate their achievements through writing. As a result, contributors to this volume are engaged primarily with developing the critical language required to address television performance. There are, of course, precedents for such an endeavour. In asking the question 'what do actors do when they act?', John Caughie's contribution to Jonathan Bignell, Stephen Lacey and Madeleine MacMurraugh-Kavanagh's book *British Television Drama: Past, Present and Future* (2000) centralises the significance of the actor's work, setting up his argument by highlighting acting as 'one of the particular pleasures of British television drama' (Caughie 2000: 162). Although it did not immediately lead to a profusion of work on the television actor, Caughie's chapter has served as a catalyst in more recent work on television acting, prompting Tom Cantrell and Christopher Hogg's concern that there is 'still significant ground to cover, particularly in investigating the ways in which actors mobilize and adapt their training and techniques of preparation and delivery to meet the demands of televisual storytelling' (2016: 2). While many of the chapters in this collection, and especially that of Gary Cassidy and Simone Knox, are concerned with what actors do (and their processes), the preference for using 'performance' as an overarching term in this collection rather than 'acting' points to a desire to widen the scope of critical discussion. This is not least in response to the breadth of television programming, in which not every character or person we see on-screen is an actor in a traditional or professionally defined sense. If acting is 'what the actor does in front of the camera' (Cantrell and Hogg 2016: 2), there is an implicit limitation on who and what can be discussed. Or, to put it in the clearest terms, everyone on television performs, but not everyone acts. For our purposes, performance then refers to what anyone (in this collection, human or non-human) does in front of the camera, as well as to how that work is organised and constructed for the audience in terms of the performer's relationship to other elements of audiovisual style (lighting, costuming, editing and so on) and to their broader integration into the demands of the programme and its contexts (production, genre, seriality, format). As our key term, 'performance' is deliberately expansive, illustrating an investment in situating the work of individuals within the complex achievements of surrounding production contexts.

In keeping with this effort to broaden an understanding of what television performance might be, this edited collection aims to keep a range of different critical approaches in play rather than settling too quickly upon one defining method. As the contributors respond to the eclectic range of opportunities available for discussing and appreciating performance in television, their chapters also help to reconceptualise notions of performance itself. This, necessarily, is a process of complication that might frustrate a desire for singular, straightforward definition but is, we would contend, best accommodated and expressed within the form of the edited collection; each essay confronting the disciplinary question through case studies, deploying a variety of critical techniques, theoretical tools and methodological interests. In order to guide readers through these approaches, we have grouped chapters into thematic sections. In Part 1, we have included those whose analysis of performance engages most directly with the form of television, with chapters examining the particular challenges of assessing performance in the dramatic episodic series (Chapter 1 by Sarah Cardwell); the provocative example of things and animals, whose performances are part of a bigger labour of production and aesthetic achievement (Chapter 2 by Jonathan Bignell); the capacity of television to keep performers in the same roles over a number of years or even decades (Chapter 3 by James Walters); and the interpretive challenges in approaching the unfolding performance in serial drama (Chapter 4 by Elliott Logan). In Part 2, we feature chapters that contend with groups of performers in a range of different television genres, including soap opera (Chapter 5 by James Zborowski), sitcom (Chapter 6 by Tom Brown), structured reality (Chapter 7 by Amy Holdsworth and Karen Lury) and panel show (Chapter 8 by Alex Clayton); these engage with the differing forms, cohesions and disruptions of interaction, sociability, play and antagonism that ensemble performance entails. In Part 3, chapters focus predominantly on the achievements of single performers, with each reflecting the different possible modes of approach to the individual, combining interviews with interpretation of expressive technique (Chapter 9 by Gary Cassidy and Simone Knox); the textuality of a performer's work across serialised television (Chapter 10 by Lucy Fife Donaldson); an appreciation of an actor's tonal complexity and contribution to an ensemble cast (Chapter 11 by Lydia Buckingham); and a material mapping of facial expressions (Chapter 12 by Timotheus Vermeulen).

While many of the contributors focus their attention on performances by professional actors, certain chapters feature sustained attention to the work of non-actors in television. Alex Clayton, in Chapter 8, writes about the comedy panel show, a genre which involves performers from a variety of training backgrounds and other televisual contexts: the chosen case study, *Would I Lie to You?* (BBC, 2007–), combines stand-up comedians, comic actors and a reality TV star, for example. Amy Holdsworth and Karen Lury's study of crying on television (Chapter 7) is occupied primarily with reality programming which accommodates multiple registers of performativity, albeit all from non-actors. This ranges from shows that most overtly display performance – *The X Factor* (ITV, 2004–; Fox 2011–13), a programme structured around amateurs singing live and directly to their audience – to those that are at pains to puncture the display of the celebrity persona and offer its subjects at their most apparently authentic – *Who Do You Think You Are?* (BBC, 2004–), which centres on the revelation of a famous personality's family history. Indeed, the possibilities and limitations of presenting authenticity on television are at the heart of the chapter's argument. Holdsworth and Lury's work places *America's Next Top Model* (UPN, 2003–06; The CW, 2006–15; VH1, 2016–) at the centre of its debates and so might be seen to incorporate the emphasis on stardom and celebrity influenced by Richard Dyer's founding contribution *Stars* (1979)[3] and continuing in the work of scholars like James Bennett (2010), Su Holmes and Sean Redmond (2006, 2007), and dedicated journals like *Celebrity Studies*. Moving even further from the non-actor, Jonathan Bignell (Chapter 2) pursues a different focus, away from human beings almost entirely, as he considers how the behaviours of objects and animals in television might be evaluated and understood as performances. In so doing, Bignell reflects on the production of television, using performance in its broadest scope to incorporate the labours of puppeteers and the material constitution of television's audiovisual style. These divergent interests represented across the book illustrate usefully the broad span of performance on television, providing our contributors with the opportunity to choose as their focus the harnessing of actor training to achieve subtle yet profound effects, for example, or technical accomplishments inherent in the manipulation of props and puppets, or alternatively the carefully composed (or distraught) behaviour of participants and judges in reality TV.[4] Incorporating this expanse, which simultaneously celebrates the value of an aesthetic approach, is what we find to be at stake in advancing the work of appreciating performance in television. The questions that drive all the chapters in this book, and their choices of subject, are: (1) What are the opportunities for performance that the specificities of television offer? and (2) What might we value in television performance?

Taking performance as our subject invites the chapters to push at wider questions, such as the pleasures to be had when professional control slips, which features in Tom Brown's discussion of corpsing in *Curb Your Enthusiasm*

(HBO, 2000–) (a term that recurs across the collection, itself prompting the thought that television might offer greater possibilities in the gaps between notions of success and failure, and for play, than those perhaps more rigidly held in film, for example). The adoption of 'performance' as the key critical term also enables our contributors to explore and complicate notions of what makes a 'successful' performance on television. This might be a value associated with a kind of training that strives to make technique invisible – found, for example, in the work of Tony Curran, an actor whose methods are made visible by the careful scrutiny of Cassidy and Knox[5] in Chapter 9. However, there are other instances of meaningful performance that do not adhere to these boundaries (as explored by Alex Clayton and Tom Brown), as well as those that allow room to consider deficiencies, even when these very qualities of skill and training are present, as James Zborowski finds in the limits of *Coronation Street*'s (Granada, 1960–2006; ITV 2006–) handling of a multi-character argument (Chapter 5). Consideration of performance can thus become a tool to educate ourselves about the variety of what people, not only actors, do on-screen and, furthermore, to reflect on the ways in which we respond. The oft-cited qualities of the democracy and intimacy of television enable a familiarity with certain performers, encouraging us to consider their central role in securing and shaping our involvement in television programming.[6] Moreover, in recognising the access this allows to such a variety of performers and their work, following the writing of those who see performance as part of social and everyday behaviours (such as Erving Goffman and Judith Butler),[7] we would argue that performance is a central conduit in moulding our understanding or judgements about people's behaviours more broadly. Coming back to the *Key and Peele* sketch, the joke that *both* men transfer their anxieties about seeming not black enough in front of another African American man into a stereotyped 'performance' of blackness, which also implicates an association of a more aggressively masculine attitude (one that does not include enjoying theatre or speaking in high-pitched voices), highlights exactly the kind of role that performance plays in everyday life. That the constructedness of such behaviour is formed through a passing mutual occupation of a purely transitional space, their shifts in gesture and speech occurring just as long as they are within each other's orbit, heightens the fluctuating dynamics in order to fully reveal them and render visible not only the artifice of such roles, but also the deftness of the performers executing them. Such attention to performance thus enables us to recognise, evaluate and even celebrate specific ways of *being* in the world. As emphasised by the editors of a special issue of the *Journal of Popular Television*, which focuses on Northern Stars in British television, it is specifically attention to performance that, in this case, 'enables us to uncover an alternative and rich thread of regionally constructed representations of femaleness and femininity,' for example (Forrest and Johnson 2016a: 196).

Performance is not only tied to the *who*, but also embraces the relationships between performer and programme, capturing a desire to attend to *how* it is shaped by its organisation in television. The writers featured in this collection are interested therefore in not just what the person in front of the camera does, but how their actions or words are shaped for an audience within framing structures such as audiovisual style, narrative, genre and reception.

Medium Specificity, or What Does the Critic Do with Television Performance?

The connection between achievement in and appreciation of performance, in any medium, is frequently plagued by the difficulties of pinning down what performers do. While it feels straightforward to identify the transformations afforded by voice, face, posture and costume in Key and Peele's 'East/West College Bowl' sketch, for example, the specific choices made in the interstitial moments – especially those in which Key and Peele are both sat in a car, holding conversations while apparently driving through the desert – are much trickier to identify and convey. The challenge to critical analysis is a central preoccupation for Caughie, who concludes his chapter by suggesting that acting's trickiness to convey may be seen as a positive: 'what exactly actors do when they act may elude the vocabulary of critical analysis, but the elusiveness only makes the question more interesting to ask' (2000: 170). This hard-to-grasp quality is frequently cited in work on film acting and performance, a more established field from which writing on television can inevitably draw. It is not always desirable or useful to look towards film studies as a way of measuring the shape and development of television studies, as though one discipline were following perennially in the footsteps of the other or, indeed, forever walking in its shadow. Nonetheless, the growth of performance as a critical concern in film studies does offer some potential areas of useful resonance for television studies and key writers, such as James Naremore (1988) and Andrew Klevan (2005), certainly inform the work of writers in this collection. And yet each chapter seeks to embrace the specific challenges of televisual performance, which are not limited to the difficulties of description or comprehension. Indeed, the particularities of the medium in both production and form offer further obstacles to and rewards for approaching this topic, thus requiring specific interventions into critical language to be used as a means of appreciating the achievements of television performance.

The way television is made challenges both actor and critic, entailing the adoption of specific approaches for each. The reduced or absent rehearsal time of television features as a consideration for Zborowski's attention in Chapter 5 to soap opera acting, for example, prompting the adoption of an analytical toolkit borrowed from non-fiction broadcasting as opposed to the perhaps more obvious contexts of theatre or film. Chapter 9 by Cassidy and Knox incorporates interviews conducted with the actor Tony Curran as a means of underpinning critical analysis. In doing this, their contribution connects with a methodology adopted by Máire Messenger Davies and Roberta Pearson (Pearson 2010: 166–83) in their interviews with Patrick Stewart and other *Star Trek* production professionals, and expanded into book-length studies by Richard Hewett (2017), and by Tom Cantrell and Christopher Hogg (2017) who build a persuasive account of acting in British television based upon a series of interviews conducted for each chapter. In adopting this methodology, Cassidy and Knox make a case for the emotional labour of television performance, a dynamic which usefully contributes to our conceptions of what it is that actors do and how they sustain their careers in doing it, thus helping to expand the scope for consideration beyond that of the programmes themselves. Extending this reach might allow us to connect with and reflect further upon issues of control and ownership across the case studies in other chapters, from the professional accomplishment of inhabiting a long-running character, evident in the tightly controlled performance of Lena Headey in *Game of Thrones* (HBO, 2011–), as illuminated by Walters' analysis in Chapter 3, or the absence of control for the reality contestant, despite their efforts to alter this through emotional manipulation/authenticity, as elucidated by Holdsworth and Lury in Chapter 7. Similarly, Clayton in Chapter 8 highlights spontaneity and improvisation as distinctively televisual aspects of performance and, in her appreciation of Jennifer Aniston's work in the sitcom *Friends* (NBC, 1994–2004), Lydia Buckingham appraises the ways in which acting in front of a live studio audience influences profoundly the actor's capacity for tonal responsiveness.

In terms of form, one of the greatest challenges that television provides to both performer and critic can be found in its temporality, with programmes of all kinds broadcast over a spread of years or, certainly, lengthy viewing times. Performance analysis is therefore complicated by the sheer volume of work, and specific strategies can be required to address this. Chapter 10 by Lucy Fife Donaldson, on the intertextuality of Timothy Olyphant's work in *Deadwood* (HBO, 2004–06) and *Justified* (FX, 2010–15), takes this difficulty as a primary interest in order to illustrate and to interrogate the usefulness of the moment in analysis which aims to understand the richness of television performance. Elliott Logan's reflections in Chapter 4 on *Homeland* (Showtime, 2011–) deal with the temporal dimensions of seriality, in particular the provisionality of meaning in an unfolding dramatic event. The more extreme cases offer yet further opportunities to test the specific demands of television on the performer, as explored by Walters' chapter, which attends to performances that must remain consistent while offering textural depth and richness, such as those of June Brown and Peter Falk, both of whom played one character for over 30 years (in *EastEnders* (BBC, 1985–) and *Columbo* (NBC, 1968–78; ABC, 1989–2003) respectively).

In dealing with television performance, the critic is always left with the possibility of multiple uneven fragments rather than neatly patterned moments, and so any detailed response requires a handling and manipulating of the shape of the performance/programme while attempting to respond to it. This reshaping and remaking of television in order to bring out the elements that are enticing and significant is productively explored in Chapter 12 by Timotheus Vermeulen, which redraws the parameters for analysing performance by offering an exploration of facial expressions as landscapes. These interests reflect an attitude shared among all the authors in this collection as they strive to emphasise the relationship between the work of the television professional and that of the critical viewer in the interpretation of performance and the creation of meaning. Each chapter resists elevating either group as an authority over the other, however, instead giving weight to those processes of interaction and engagement that exist between performers and their audiences. The work of the critical viewer is central to this project: reaching for the right words to convey what the performer does and how we might understand that work and, in doing so, recognising the multiplicity and possibilities of television performance. Furthermore, we might acknowledge that it is the underlying responsibility of such criticism to respond to the particularities of performances on television with sensitivity and rigour. This point is afforded extended consideration in Chapter 1 as Sarah Cardwell attends to the achievements of episodic television fiction, which might otherwise be missed if evaluated at surface level or according to a set of preconceived value judgements.

Performance Types

Breadth of performance type is related to breadth of programmes in this collection, with chapters encompassing a wide range of styles, tones and genres. This is something special to television as a medium, which offers an array of content unrivalled by cinema, for example. US 'quality' television has often, somewhat unhelpfully, been described as 'cinematic' (Jaramillo 2013: 67–75) and, although those kinds of definitions are largely avoided, precisely because of the ways the collection is responding to performance, the merits of US comedy and drama are certainly explored in chapters by Brown (*Curb Your Enthusiasm*, Chapter 6), Donaldson (*Deadwood* and *Justified*, Chapter 10), Logan (*Homeland*, Chapter 4), Buckingham (*Friends*, Chapter 11), Vermeulen (*Friday Night Lights* (NBC, 2008–11), Chapter 12) and Walters (*Game of Thrones* and *Breaking Bad* (AMC, 2008–13), Chapter 3). In some ways, this emphasis mirrors the attention that is often afforded to these kinds of shows in awards ceremonies and industry promotion, where a celebration of performance can often feature strongly. It is also the case, however, that even those things that are well known and well loved can remain critically underexplored, and each of the chapters serves to illustrate the value in looking again at the familiar and the acclaimed in order to articulate accomplishments in closer, more precise detail. But these kinds of examples are also balanced against shows that can be more neglected, not only in terms of perceived quality but also in relation to performance. Chapter 1 by Cardwell on *Death in Paradise* (BBC, 2011–), for example, brings to mind the notion of 'invisible television' as defined by Brett Mills, whereby 'There is television that gets watched and there is television that gets discussed: the two do not necessarily coincide' (2010: 1). In this respect, we might profitably place Cardwell's chapter alongside Clayton's on the panel show *Would I Lie to You?* (Chapter 8) and even Zborowski's on *Coronation Street* (Chapter 5). In each of these cases, there is the risk that performance achievement might get missed entirely, either through perceived textual blandness (*Death in Paradise*), repetition and intimacy (*Coronation Street*) or because it features within a ubiquitous or straightforward show format (*Would I Lie to You?*).

Appreciating Performance

Ultimately, the chapters in this collection all share a commitment to analysing performance in detail, using textual evidence as a means of shaping evaluative conclusions. This approach might represent a broader effort to pay greater attention to performance more generally and, as a consequence, acknowledge a lack that has hitherto existed in television studies but has been filled with greater consistency elsewhere. Andrew Klevan's subtitle for his book on film performance, 'From Achievement to Appreciation', draws a direct and symbiotic relationship between the work of the performer and the work of the critical viewer, for example. Although taking time to emphasise the performer's relationship to a wide range of compositional features within films, he is nevertheless careful to keep the individual's achievements at the heart of debates. So, for example, in Klevan's account the pivotal entrance of the character Norma in *There's Always Tomorrow* (Douglas Sirk, 1956) 'depends upon [Barbara] Stanwyck managing the relationship between intimacy and distance: her character remains open and invites trust, while withholding information about herself and her motivations' (2005: 59). Similarly, for all our writers, whatever their approach and subject, what the performer does (whether they are a reality contestant, seasoned television actor or even a rubber ball) is responded to with a detailed appreciation of their contribution to the whole. The sensitivity Klevan shows to an appreciation of the 'complexity of a performer's internal relationships within a film' (2005: iv) offers a rich model for engaging with both the details of what performers do and how their work is shaped by the medium.

Related to this, when concentrating on performers whose work has already received significant audience and industry recognition, such as Jennifer Aniston and Damian Lewis, or programmes that are celebrated in various ways such as *Coronation Street* or *Curb Your Enthusiasm*, our writers resist the temptation to allow an existing positive consensus to restrict opportunities for closer scrutiny and debate. Each chapter works from the perspective that there is still more to say about work that has already been admired and acclaimed. Equally, chapters find space to discuss in detail performances that don't necessarily 'stand out' yet nevertheless complement the shape and tone of the programmes in which they occur. Both the remarkable and the exemplary are worth investigating, as those performances singled out for being especially brilliant or complex in their achievements (as with Jennifer Aniston or Damian Lewis) are as valuable as those that are seemingly less remarkable (Tony Curran, Cheryl Hines) or that are representative of ways in which performance on television might work more broadly (the casts of *Coronation Street* or *American Idol*) to the overall picture of the expressive place of performance in television studies. Timothy Olyphant's performances in *Deadwood* and *Justified* require further thought because their ostensible similarity, while in tune with each show's dramatic texture, presents certain challenges for claims of individual quality. Likewise, when a lead role is replaced in a programme, as with *Death in Paradise*, the work of the incoming actor may get lost as it is shaped not only by the legacy of the predecessor but also by a need to maintain a narrative coherence that has already been developed. Again, closer examination can yield rewards.

As the collection invests in a broad range of different performance styles, it will become evident that no single criterion for achievement is insisted upon. Indeed, through shared processes of detailed analysis, of close engagement with a performer or performers, the claims of our authors for specific performance qualities respond to the shape and structure of the programmes themselves. The choice of case study by each author is carefully selected to answer the specifics of their argument. By implication, then, a different selection of performances and programmes might take the discussion in new directions and towards other conclusions. Here, both the reach and limits of the collection are made clear. The selection of texts represented in these 12 chapters extends an appreciation of performance in television but at the same time points towards fresh horizons for debate that exist beyond its pages. One such area of expansion could be the application of this book's debates across other national contexts. The global television industry means that US and UK shows are exported widely to substantial territories like China and India as well as mainland Europe. Those rights deals are often sewn up for shows like *Game of Thrones* and *Homeland* before they are broadcast, while streaming on Netflix and other online platforms further increases the potential for access (depending, of course, upon each nation state's attitude towards censorship and restriction). So, in this context, viewers around the world are very familiar with US and UK shows and this collection's focus on performances deriving from these contexts is intended precisely to provide a viable touchstone for the topic of television performance. This resonates with the book's overt aim to offer a representative study of performance rather than of the global television industry. One of our principal aims, of course, is that readers will be able to apply the debates found in the book to other examples and to compare our choices of focus with others. In this respect, it is hoped that the collection's value will be found in its capacity to open up and sustain the field of television performance as it becomes a more prominent area within television studies.

Self-evidently, this collection is not intended to be a final word on television performance. Rather, it is an invitation for the reader to take the conversation further, to measure the work of our authors against the programmes themselves, and to engage in a continuing, detailed understanding of television performers, their achievements and their value.

Notes

1 These moments in the car feed into a wider parody of the first season of *True Detective* (HBO, 2014) which includes the opening credits.
2 Directors do not enjoy anything like the kind of recognition in television that they do in film, and remain largely absent in promotion, audience reception, and critical response. The notion of director as author of meaning has not taken root in television and, indeed, directors are most likely to be recognised if they have a previous association with cinema (Jane Campion, David Fincher, Ridley Scott and David Lynch, for example).
3 Followed by a continually growing field of scholarship, as indicated by the BFI Stars Series.
4 There are further figures that might be brought into this conversation through an expanded idea of performance, such as the stunt double, a member of the production team kept deliberately invisible in the work of performance but crucial to its success, as indicated by Zoë Shacklock's article on acting doubles in *Orphan Black* (Shacklock 2016).
5 This critical dynamic of technique and careful attention to achievement is a key focus of Cassidy and Knox's blog series for *CST Online*, 'What Actors Do', which has provided a valuable early contribution to scholarship on television performance.

6 Elliott Logan writes about performance and involvement in his article 'How Do We Write about Performance in Serial Television?', which opens with a consideration of 'the way our involvement in the best serial television is strongly keyed to the presence of particular performers over the periods and rhythms of time available to such shows' (Logan 2015: 28).

7 See: Erving Goffman's *The Presentation of Self in Everyday Life* (1959) and Judith Butler's *Gender Trouble* (1990).

Part 1

Performance and Television Form

1

In Small Packages: Particularities of Performance in Dramatic Episodic Series

Sarah Cardwell

The accumulated amount of character knowledge that a long-term serial viewer accumulates gives a richness of interpretation to any single instalment.

(Smith 2014: 290)

The familiarity with characters that is furnished by prior episodes is sustained and built upon through the incorporation of gestures, moments, and events that are imbued with emotional weight based on information parcelled out previously.

(Nannicelli 2016: 70)

As the editors of this volume observe in the Introduction, attention to performance on television has, until recently, been relatively limited; this collection constitutes a welcome and significant expansion. Within existing scholarship, some of the most incisive and nuanced accounts of television performance are proffered in the context of focused analyses of specific programmes and characters, under the aegis of television aesthetics (which is also where this chapter situates itself).[1] One of the most important, obvious and immediately engaging functions of performance is, after all, to express and communicate the development of character. Thus valuable insights into televisual performance have arisen via critical and scholarly attention, such as that paid by Greg M. Smith and Ted Nannicelli as highlighted in the above quotes, to notable examples of extended character development in TV serials such as *Mad Men* and *Breaking Bad*.[2]

Notably, these are both long-running dramatic serials, proffering complex narratives stretched out over many series/seasons. Recent expressive and evaluative criticism of television frequently valorises TV's seriality and the dramatic potential of long-running serial form, especially its manifestation in 'quality US drama'. The episodic series, also a firmly established, persistent televisual form, composed of fully self-contained episodes ('small packages') in terms of plot, and dependent upon core characters supported and supplemented each week by changing cameos, lies comparatively neglected by aestheticians and cognitivists, and undervalued by critics.

The widespread preoccupation with serials is understandable, reflecting a desire to pinpoint some of the specificities of TV, its dominant forms and noteworthy achievements. In their collection on television aesthetics and style, Jason Jacobs and Steven Peacock identify extensive seriality as a primary difference between television and film, citing 'the expansive structure of television fictions' (2013: 6–7). Ted Nannicelli, in his philosophical account of the art of television, explores 'temporal prolongation' in his endeavour to establish the specificity of TV: 'the television medium possesses certain qualities, including the capacities for temporal prolongation and liveness, that differentiate it from the film medium' (2016: 80, 81). Others, such as Robin Nelson (2007) and Jason Mittell (2015), have commended contemporary 'quality' TV serials' complex narrative and character development.

Furthermore, it is easy to see why television aestheticians, who seek particularly to pinpoint the singular achievements of notable works, might tend less often to be inspired by programmes that exhibit formula, repetition, conformity, lack of individuality. While episodic formats garner considerable attention from mainstream TV studies, within television aesthetics – which promises detailed attention to elements such as style and performance – serialised dramas are far more often the focus of recent close analyses than are episodic ones. The fact that episodic series are frequently generic – as in the case of *Death in Paradise* (BBC, 2011–), the police procedural which is my chosen focus – rather than 'serious' drama, compounds the problem.[3] Scholars concerned with details of character and performance are understandably drawn to explore and appreciate the achievements of notable serials which allow for the extensive, intricate development of character.

From a distinct but potentially complementary perspective, nascent work undertaken by cognitivists interested in television adds weight to the focus on seriality. Cognitive scholars begin from a reasonable presumption that our engagement with characters is central to our relationship with and appreciation of the work in front of us, and that, moreover, moving-image narrative forms such as film and television exploit instinctive, deep-rooted ways in which we form real-life connections. It is clear, therefore, why the rise in profile of long-running serials has lured some cognitivists from their home ground of film studies to television studies. Television serials facilitate far more obviously than do films the 'friendship metaphor' as a model for our complex engagement with on-screen characters: cognitivists Robert Blanchet and Margrethe Bruun Vaage argue that 'by generating an impression of a shared history, television series activate mental mechanisms similar to those activated by friendship in real life' (2012: 18).[4] As noted in the quotations that open this chapter, our memories of characters' histories, accreted across the extended narrative arc of a serial, intensify our engagement with those characters. Cognitivists exhibit enthusiasm for the same long-form serials as do TV aestheticians, favouring them over episodic programmes, since the former appear to correspond more closely with their scholarly preoccupations.

Both aesthetic and cognitive approaches contend, with differing degrees of explicitness, that sophisticated characterisation and performance enable us to build intimacy with a character over a sustained period, and that this process is central to our engagement and appreciation. Thus across two key areas of television studies which concern themselves particularly with characters and performance, long-form serials have dominated recent discussion.

When performance is evaluated within this context (of long-run serials), an inevitable focus emerges on the profound exploration or long-term progression of characters, and the actors who develop with and through them. However, this is not the only possible route into television performance, nor does it cover the range of pleasures to be found in particular performances, especially where they exist in non-serial episodic forms and genres. This chapter addresses performance within conventional episodic form. Episodic form can be arguably more challenging for its creators and performers, given its restrictions and limitations. Herein, I recognise and explore some of the implications of episodic form for performance, and celebrate the respective achievements of one specific, oft-derided series, *Death in Paradise*. Within necessarily limited space, I hope to show that it is possible to proffer sensitive and appreciative explorations of particularities of performance within traditional episodic series. Underlying this chapter is a firm belief that sometimes good things come in small packages.

Counterpointing the Serial with the Episodic

The current dominance of serial form within television scholarship and criticism raises two potential barriers to the appreciation of performance(s) within episodic series.

First, important though the serial form is to contemporary quality TV, there is a risk that complex seriality begins to stand for the fully realised 'televisual', or at least for that worthy of close aesthetic attention; our understanding and appreciation of serial form consequently belies a lack of responsiveness to television's other dramatic traditions and to the particularities of performance therein.

There is a tendency not only to acclaim recent long-running serials but also to proffer, if only for classificatory or rhetorical purposes, episodic form as their logical counterpoint. Some scholarship, while celebrating the open, evolving serial, comes close to caricaturing the episodic series as a dead counterweight: closed, constrained, simplistic, out of date. Bruun Vaage contrasts 'regular TV' as LOB (Least Objectionable Programming) for an 'undifferentiated mass audience' against newer, long-form US quality serials (2016: xii). Mittell's work on 'complex' contemporary US television offers probably the most sustained deployment of this use of the episodic series as counterpoint. In one essay, extolling the virtues of recent US television serial dramas, he depicts episodic form as historical, and claims that narrative complexity develops in television only from the late 1990s; aligning the latter with serial form, Mittell argues that this 'new TV is more "difficult"' than earlier or other television (2010: 78, 79).

It must be acknowledged that most of the scholars cited above offer caveats regarding their preference for serials over episodic series. While Mittell allies complexity to serial form, he also clarifies in a later work that complexity is not necessarily to be equated with value, and that simplicity is sometimes artistically preferable (2015: 217). Jacobs and Peacock (2013) are careful to emphasise that neither quality nor value resides inherently within any particular form of TV, and Nannicelli asserts that there is no 'direct, causal link between temporal prolongation and artistic achievement' (2016: 81). In relation to close analyses of specific instances of television, Jacobs and Peacock further advocate an approach which explores the relation of a singular 'moment' to the whole, while Nannicelli offers a precise and nuanced account of the relationship between individual episodes and the entire serial/series (2016: 108–13). Thus in these scholarly accounts, the facet of seriality does not dominate, or undermine attentiveness to, the particular (however the latter is defined).

Nevertheless, the widespread prevalence of serials as examples, and the not infrequent use of juxtaposition (counterpoint and contrast) to highlight the value of extensive seriality compared with episodic form, tend to divert attention away from the particularities and achievements of episodic dramas, even though many of the ideas presented in relation to serials – and relevant to explorations of character and performance – apply with equal persuasiveness to episodic series.[5] Enthusiasm for the recent long-form serial has unintentionally discouraged close attention to the specific qualities and achievements of episodic series; correspondingly, consideration of performance within dramatic episodic series lags behind.

Second, the overemphasis on serials impacts particularly upon the study of character and performance. Valuable analyses of performances within serials inadvertently begin to institute implicit notions of what successful, persuasive performances look like. The element of seriality leads us to value change, development, or at least deepening interrogation or uncovering of character. Performance contributes to narrative and/or character development: there is a sense of movement, progression, going somewhere. In these terms, the contrapuntal episodic series appears to go nowhere – and the same could be said of its characters who, especially in procedurals, must exhibit constancy, repetition, perhaps circularity. Tacit assumptions about successful performance thus risk leading us to a dead end when attempting to appreciate character and performance in the near-amnesiac episodic series, wherein character development is necessarily severely limited and each episode must be self-contained, plausibly absorbing any change, complication and resolutions.

Episodic Form and Character Development, Revelation and Performance

If we seek, above all, complex and elongated diegetic development as the epitome of successful characterisation and performance, we can only be disappointed by the episodic series.[6] The conventional episodic series does not generally allow for notable change or progression. However, it extends us the pleasure of repeated confirmation of character, and it can proffer an elucidation of character in increasing detail, enabling us to get to know a character more intimately each week, deepening our rapport, regardless of the presence, length, complexity or profundity of that character's individual journey. While the form is amnesiac in terms of plot, it need not be so in terms of character: a nuanced performance that is straightforwardly accessible to new viewers can simultaneously allow fresh insights into a character's disposition that will be detected and appreciated by the regular audience.

Scholars and teachers of acting have noted the potential continuity between screen performances and social relationships. Patrick Tucker, in his practical guide to screen acting, avers 'We are all stage actors' (1994: 3), in the sense that our words and bodies present a performance that others may interpret as expressive of our characters, thoughts and feelings. It is through how others act that we come to know and appreciate them. Importantly for our purposes here, what matters most to us is a person's recurring traits: the details of how they *habitually* behave, speak, move.

Though cognitivists appear in thrall to serial form, it is striking that in terms of performance there are clearer correspondences between our engagement with characters in *episodic* form and with others in real life. How often, after all, do we scrutinise and invest in the individual, life trajectories of those around us, their personal 'development', in the same way as when we watch a long-form serial? We can befriend someone over a decade without having privileged access to his or her 'narrative', inner life, intimate relationships and memories. Instead, as we get to know someone better, we may take pleasure in their very consistency: in the reiteration of words and actions that typify the person. Over time, in some cases, we might deepen our understanding and appreciation of their character via our observations, and this can constitute a significant and pleasurable relationship even if the person undergoes no striking changes or development.[7] In short, we get to know those around us well precisely because of repetition and reliability. In this way, an appreciation of the kind of performances found in episodic series might chime with our appreciation of other people in real life.[8]

Thus in the episodic series, enjoyment and satisfaction can be sought in performances which particularly exploit our common experiences of friendship and acquaintanceship in the real world. If we shift our focus from the extensive development of character, and from the contribution of performance and character to narrative progression and complexity, we can attend instead to the ongoing pleasures of the revelation, shaping and affirmation of character via accreted details of performance. (Indeed, we can also relish those details for their own sake, in the moment, even if they prove evanescent, adding nothing noteworthy to our perception of the character or any broader narrative arc.)

Those details of performance will have their own specificity. As James Naremore neatly defines it, acting is the 'systematic ostentatious depiction of character' (1988: 23); we must recognise that the context or 'system' differs between serial and episodic forms. Successful performances in episodic series must effectively deploy the essential trope of repetition (present also in serials, but more marked in episodic series), while avoiding tired repetitiousness, and delicately balance several things, including forward momentum with equilibrium, and character revelation with the need for immediate recognisability, all the time avoiding caricature.

There are other important facets to be considered when attending to a performance within episodic form. Most such series are patently ensemble pieces, and the ensemble is a primary source of delight. Therefore performances are likely to be interdependent, each actor/character allowing others to play their part. Relationships between characters must exhibit constancies and continuities to allow the programme's format to persist, and any changes must be absorbed without undermining the entire equilibrium of the work. *Death in Paradise*, for example, has faced several times the departure of key actors, whose roles and functions must be thoughtfully and plausibly replicated.

Even more fundamentally, the episodic series requires that we trouble the simplistic connection between performance and character. Performance in the episodic series fulfils additional functions and offers alternative satisfactions. *Death in Paradise* exploits repetition as a source of pleasure, as repetitive motifs of performance are central to the programme's broadly comedic tone.

Death in Paradise: Detective Inspector Richard Poole

Death in Paradise is ignored by scholars and sneered at by critics, but loved by audiences. This light-hearted police procedural is conspicuously conventionally structured – more so than some other recent examples (such as *Castle* (ABC, 2009–16) or *The Doctor Blake Mysteries* (ABC, 2013–17;

Seven Network, 2018)), which incorporate recurrent back stories and protracted relationships between characters. *Death in Paradise*, defiantly old-school, embraces episodic form wholeheartedly. Each episode is entirely self-contained – the complete package[9] – following a predictable narrative format, and comprising a central, constant cast complemented by weekly cameos.

The programme is set on Saint Marie, a fictional Caribbean island; it is filmed on location in Guadeloupe. The island is a star in its own right: *Death in Paradise* is shrewdly scheduled in British winter, allowing viewers to savour the stunning scenery, sandy beaches, vibrant bars and Caribbean music. Indeed, the programme's unashamed attractiveness is likely one of the reasons for critics' scorn: this is the very definition of escapist television.[10]

In the opening episode of series one, DI Richard Poole, played by Ben Miller, arrives from cold, damp London, a fish out of water, to oversee the local police force, who have found themselves in the unusual situation of having to investigate the murder of their Detective Inspector, Charlie Hulme (Hugo Speer). This episode establishes the programme's format, with pre-credit scenes showing the lead up to and discovery of that week's murder case: here, Inspector Hulme is found shot dead in a safe room, in a classic 'locked room mystery'. The investigation is undertaken by a small stalwart team, the show's regular cast: laid-back Dwayne Myers (Danny John-Jules), keen novice Fidel Best (Gary Carr), and astute DS Camille Bordey (Sara Martins); 'The Commissioner' (Selywn Patterson, played by Don Warrington) also makes periodic appearances.

The series' credit sequence – a cheery, golden-hued montage of the central characters, idyllic beaches and palm trees of Saint Marie – includes shots of DI Poole already established among his colleagues. However, our properly diegetic introduction to the new inspector is one of gradual exposure via a series of fragments. Our first glimpse of the man who is to step into the shoes of the recently deceased DI is, appropriately enough, of his feet – or at least, his lower legs, from his shiny, lace-up brogues, which gleam in the light, up to mid-calf height, his legs clad in a well-cut, conservative dark blue suit. Poole's long, swift strides through the arrivals area of Saint Marie airport are purposeful and confident, and his attention to detail and tendency to formality resonate in his brusque yet polite, enunciated middle-class Southern English tones, as he proclaims, with barely concealed impatience, that the airline has lost his luggage. His idiomatic English phrasing of a request (or order) as an incomplete question, 'If you could just point me in the direction of the lost luggage desk', combines politeness and authoritative firmness in equal measure. Poole's request coincides with a sharp about-turn to face his interlocutors: he wheels around on his standing leg, drawing his free foot to close neatly in an almost military, almost balletic movement that expresses neatness, precision, decisiveness. Poole is given the foreground: the toes of the other characters can just be seen in the distance, but the outline of Poole's feet and legs is sharply drawn (Figure 1.1).

At the helpdesk, as Poole fills out a form regarding his missing suitcase, the focus of the shot moves to his mid-section, arms and hands. He asks what time the lost luggage desk opens and, on being informed that it is at 6 am, replies, 'Then I'll call you at oh-six oh-one, thank you very much', with rising and levelling upward inflection, already certain of his interlocutor's compliance. Notably, Poole extends his arm and shakes back his sleeve to check his watch in the moment before he enunciates '06:01'. This gesture is in practical terms redundant, but is expressive and indicative, establishing that Poole is prone to emphatic, self-assuring gestures to underscore his assertions, and does not shy from performing for those around him. This

Figure 1.1   DI Poole's feet (*Death in Paradise*, BBC, 2011–)

personal quality lends plausibility to the later establishment of another key, generic element of the programme's format: the denouement in which the inspector verbally unravels the devious machinations behind that week's case and reveals the identity of the murderer, exposing one of the gathered suspects as the guilty party.

As Poole walks out of the airport, he is filmed from behind in a head shot, his dark suit and hair punctuated by the crisp white collar of his shirt. And then, at last, we see his face, as he pauses at the airport doors and squints up into dazzling sun, muttering 'Christ'. His pale face is already pink and slightly sweaty, and he appears breathless from the searing heat. In this moment, DI Poole's dark, formal, polished attire – while still elegant – is clearly inappropriate. He is both literally and sartorially over-dressed. The tension established here between Poole's instinctively confi-dent and commanding demeanour and his discomfort within unnervingly unfamiliar surroundings is exploited throughout the series, mostly to comic effect.

Indeed, Miller's performance tends throughout towards the comedic (later inspectors move the dial further towards the dramatic).[11] And the deliber-ately fractured presentation of the actor and character in this introduction function less to build up tension about who the actor is and more to direct our attention to details of his distinctive form, physiognomy and voice; these elements of physicality constitute from the beginning major aspects of his performance.

Poole's confidence, precision, sharpness, reserve and awkwardness (traits regarded by other characters in the programme as typically 'English') are conveyed not only by what he says but also by the actor's mode of delivery. Miller's distinctively dry tone renders Poole's sarcastic observations of the many 'problems' of Saint Marie (excessively hot weather; strange food; laid-back locals; odd customs; no cutting-edge technology) amusing, tongue-in-cheek, rather than tiresome. Poole concludes his sourer comments with a perceptible pinching together of his lips, indicating tight disapproval, per-sonal repression and a rather prissy self-consciousness which encourages us to laugh with and at him simultaneously.[12]

Above all, Poole's character is manifestly embodied, concentrated within details of physical and kinaesthetic performance: facial and bodily expres-sion, posture, poise and movement. When he first arrives at the police station, set above a small, bustling market square, a mixture of close-ups, mid-shots and long shots capture the details of Miller's corporeality. Discombobulated, Poole gazes atypically open-mouthed, turning to absorb the scene. Taking control of himself, he heads towards the steps that lead up to the police station. With impressive flexibility, he surmounts the stairs with lengthy, bending strides, but this lower-body fluidity sharply contrasts with his stiff upper body, which exhibits no natural counter-body movement. Over Poole's stiffly crooked left arm hangs a superfluous mackintosh, and in his right hand is a briefcase, held unnaturally low by a rigidly straight arm. Hunching his head down into slightly curved shoulders, he looks short in the neck, as if defensively shielding himself from any unexpected blows. The overall impression is comically unnatural and awkward. But it is not merely funny: it also encapsulates Poole's keenness to get on, speediness and determination versus his stiff reserve, self-containment and lack of ease. As is often the case in these kinds of programmes, comedic performance is here balanced with and integrated within the dramatic concerns of the programme.

Strategic choices of shot size frame Miller's performance, frequently pulling back to offer a full-length shot of Poole's distinctive body movements, posture and poise, as in a sequence within this opening episode in which he makes a phone call to England from his secluded beach house. A long shot frames two sets of French doors, separated by a short dividing wall, which open onto the decked veranda with its aged wooden balustrade. Poole paces absentmindedly with awkward gait back and forth, crossing the space of one doorway, disappearing briefly and then reappearing to cross the other before reversing his trajectory, pausing intermittently to speak to his interlocutor (Figure 1.2). Knowing he is unobserved, his usual restraint is eased, and he is more expressive with his body. Concentrating on the fuzzy phone connection and absorbed in the call, he curves slightly forward and inward, a gentle 'C' shape, as if suffering from a mild but uncomfortable stomach ache, a semi-bowed posture which complements his courteous language in suggesting a measure of deference to the person on the other end of the line (his superior officer in London). Sometimes, Poole reverses the shape, rising on his toes and swaying his hips forward and his chest back, as if by stretching out his body he might find that it all falls comfortably into place. Again, Miller's movements, while entertaining, also imply his flustered emotional state; his physical and mental conditions are inextricably bound within his performance.

Figure 1.2   DI Poole on the veranda (*Death in Paradise*, BBC, 2011–)

At one point, in a sweeping gesture accompanying his claim that he has entirely settled in and is comfortable in his surroundings, he runs his hand blithely along the top of the wooden balustrade – and immediately encounters a splinter. Grimacing, but trying to conceal the problem from his interlocutor, Poole bends more deeply forward at the waist as if punched in the stomach, raising one leg and lowering his head in exaggerated movement of pain and frustration, before dropping his head and sagging his shoulders in defeat. Much of this scene is shot in long shot, and the bright sunlight streaming into the beach house silhouettes him to some degree, highlighting the shapes he forms with his body. Such details of performance and its framing/staging may be more wholly appreciated when watching slowly and attentively, but at the same time Miller's performance here is drawn with broad enough brushstrokes that anyone viewing this scene for the first time can readily engage with and appreciate his character, as well as enjoy the humour.

Miller's DI Poole is a man out of his comfort zone, and the comedic pleasure to be found in his performance arises frequently from his contrast with his surroundings. He is uncomfortable walking on sand, rarely sits with his colleagues in the local café, and makes it clear that he looks forward to returning home to England. He is also markedly distinct from his colleagues, their scepticism and exasperation providing a choric counterpoint: they act as wry observers of and commentators upon his performance. As Poole takes his seat in front of a large desktop computer, insisting it is imperative he access the network 'asap', hitting the side of the monitor and complaining that it won't turn on, laid-back local cop Dwayne, who has the air of a man who has seen it all before, observes sotto voce to his colleagues: 'This is not going to go well.'

This is not to say that Poole is merely a butt of humour. He is bright and scientifically minded, coming up with ingenious ideas to make up for the lack of technology on the island, and diligent and persistent as he pursues the smallest case details. He is gifted in a way that sets him apart from his team. His verbal proficiency is impressive and, as noted above, his English reserve does not prevent him wholly enjoying the performance that constitutes the denouement of each episode – this he does with flourish and theatricality, to the extent that in later episodes he is not ashamed to choreograph and direct other members of the team. The generically conventional, revelatory finale of episodic detective series demands from an actor a performance suitably theatrical, and yet consistent with the character as already established. Here, as in several other examples of the genre, DI Poole's self-regard for his intelligence and proficiency provides plausible motivation for him to overcome his instinctive introversion, and seize with relish an opportunity to display publicly his impressive abilities.

In the Miller episodes, then, the inspector is a protagonist whose dramatic and comedic significance depends upon contrast and juxtaposition. His idiosyncratic character traits, expressed via bold performative gestures and repetitions, are juxtaposed with the rest of the regular ensemble of characters, who retain subsidiary status. Poole's voice contrasts with their slower diction and rolling tones; his striking physical qualities and modes of movement diverge from their more relaxed demeanours and gestures. Furthermore, each actor's performance pleasurably highlights the particularities of others in the group: Dwayne's cool, laid-back, flirtatious charm; Fidel's earnest, youthful diligence and occasional naivety; Camille's feisty confidence and scepticism.

The programme constructs a hierarchy between the central characters and cameos via subtle differences in their respective performances. To take up Naremore's observation that actors must compromise between 'obviousness' and 'doing nothing' (1988: 34), the performances of cameos tend towards the former: they are more often overstated, with characters drawn necessarily quickly and baldly. The further generic necessity that each suspect must dissemble or 'perform' so that we consider as many of them as possible to be potentially guilty means that the cameo performances are in a sense already 'inauthentic'. A successful cameo performance thus constitutes a notable achievement, balancing as it must the requirement for relative theatricality with the need for plausibility: cameos cannot appear so inauthentic that they fracture entirely the episode's coherence and credibility. In contrast, the key characters may be uncomplicated in constitution, but they are plausible, rounded and we have time to get to know them – and the actors' performances are correspondingly subtler and more multifaceted, despite their apparent straightforwardness.

Thus a finely tuned balance is established by means of contrast, differentiation and juxtaposition, so that each member of the cast, whether protagonist, core or cameo, is able to take up his or her distinctive space and role within the drama. But these orchestrated balances between performers and characters, confirmed as they are by repeated traits of performance, can be upset – most noticeably when a crucial cog in the machine must be replaced with a new one. This happened when Ben Miller decided to quit the programme at the beginning of series three.

The New Inspector

The replacement of DI Richard Poole with DI Humphrey Goodman (Kris Marshall) in 2014 was the first of several big casting changes which took place after the programme's form and style, and its characters and their interrelation, had been established. Such a disturbance ran the risk of upsetting the series' internal equilibrium and alienating viewers. The creators had a tricky task to negotiate: the new inspector had to perform a comparable narrative function, allowing other characters to sustain their roles and interrelationships within the group, while also being plausibly distinct from his predecessor and opening up new possibilities. A delicate balance was needed between continuity and change. Continuity was assured via consistency of format and narrative structure, while a sense of movement, change and distinction from what came before were achieved through performance.[13]

The first episode of series three amusingly vocalises, via plausible meta-commentary, the potentially sceptical, resistant attitude of the regular viewer faced with a significant change to that which has become cosily familiar:

'But look, you weren't too keen on Inspector Poole when he came out here. We should just give him a chance – that's all I'm saying.'

'Well, he's no Chief. Not in my book [...] No one is going to take the Inspector's place.'

   (Fidel and Dwayne discuss the new inspector, DI Humphrey Goodman)

'I'm not here to take his place, you know [...] I'm not here to *be* him. I don't expect anything – any consideration. I'm just here.'

(Humphrey, to Camille.)

In these moments, other characters voice their doubts about the interloper's ability to take Richard Poole's place, and Humphrey attempts to assuage their concerns, explaining that he is not trying to 'replace' Richard but that he is there to perform a function. The same is true, of course, of Marshall.

Where the first two series employed contrast between the inspector and his exotic surroundings, primarily to comic effect, by series three the island and team are familiar elements to regular viewers. Now the new inspector is the unfamiliar element to whom we need introduction. The character, Humphrey Goodman, and the actor, Kris Marshall, must undertake a similar task: each must sensitively and deftly manoeuvre himself into his new situation in a way that confidently establishes his individuality and difference from his predecessor, without upsetting the poise of the established and successful ensemble already in existence.

The transition episode from one inspector to his replacement took a surprising turn. Despite the celebrated return of DI Poole from London at the end of series two – diegetic confirmation of his commitment to remaining on the island – the writers chose to open series three with Richard's murder: Poole is stabbed to death with an ice pick while attending a small gathering with some old university friends. His dramatic demise was a bold, potentially risky strategy for an 'escapist', easy-viewing series, with regular viewers unlikely to welcome such a stark (if temporary) change in tone.

DI Humphrey Goodman flies in from London to assume Poole's role. The nature of his introduction bears striking resemblances to that of his predecessor in series one: elements of physical performance are foregrounded, quickly establishing broad character traits and functioning to highlight, for regular viewers, his dissimilarity from Poole. Long shots emphasise Humphrey's very different physicality: he is not neat and petite but instead rather large and clumsy, with a looser sartorial silhouette of a soft linen jacket that bulges at the pockets, hangs open and flaps in the wind. Humphrey arrives at the police station by cab, struggling with luggage which includes a shoulder bag, plastic bags, a heavy, battered brown suitcase, and a dog-eared newspaper which falls periodically from under his arm, causing him to pause and juggle the other items to retrieve it (Figure 1.3). His appearance is messy and crumpled; his long fringe hangs over his eyes; he is tall, slim and gangly in his movement, tripping and stumbling as if not quite in control of all his limbs.[14] In response to the team's sceptical looks, the commissioner reassures them, sotto voce, that 'London speaks very highly of him'. In contrast with DI Poole's exaggerated performance of temporal precision ('I shall call at 06:01'), Humphrey wishes his team good morning before questioning out loud whether it is indeed still morning.

Figure 1.3   Humphrey and his luggage (*Death in Paradise*, BBC, 2011–)

In closer shots, Marshall's physiognomy and facial movements disclose his disposition. His initial nervousness is conveyed via an over-eager, toothy smile, the corners of which downturn rapidly, leaving him looking rather goofy and uncertain. Marshall's inspector is expressively open, not reserved; loose-, not tight-lipped; and ready with a smile. When thinking he furrows his brow and grimaces effortfully.

In this introductory episode, Marshall's performance exploits the kind of foregrounded comic physicality that typified Miller's inspector and, in this way, emphasises his differences from the previous character. However, the episodic series' amnesiac form means that Miller's inspector is ruthlessly forgotten by episode two, making sustained comparison needless. Once Marshall's inspector, Humphrey, is established, the use of contrast and juxtaposition which typified Miller's performance almost disappears. Humphrey seeks to fit in, to embrace local customs and experiences, to be part of a milieu, and Marshall correspondingly starts to dial down the more exaggerated aspects of his performance.

Humphrey astutely remarks to his team that he feels he has got to know the late Richard Poole a little, 'mostly by the effect he had on those around him'. Given the interdependence of performances in an ensemble setting, the replacement of a pivotal role necessarily alters the relationships between characters and actors. Notably, Humphrey demonstrates a far greater sensitivity to others than Poole exhibited: he deliberately chooses not to take his late predecessor's desk at first, instead plonking himself on a seat at a spare desk at the back of the room – a seat which promptly sinks and leaves him unseen behind his piled-up belongings. He is polite and thoughtful, and tentatively reaches out to befriend his new colleagues. Humphrey's greater engagement with those around him is manifested in a correspondingly more generous performance from Marshall, placing him on a more level footing with his peers. When other characters are speaking, Marshall's Humphrey looks from person to person, acknowledging them, where Miller's Richard tended to be the focus of the supporting actors' gaze, rarely returning it. Humphrey seeks acknowledgement and agreement from others, especially Camille. He is not commanding or bullish in the way that Poole was; he feels his way in this new environment.

The replacement of Poole with Goodman, and Miller with Marshall, alters the relationships between inspector and team, and between protagonist and ensemble cast. A new, democratic balance is forged across the ensemble, which one might have characterised as functioning previously as protagonist and chorus. Just as Humphrey embraces life on the island and embeds himself amenably within his team, Marshall's performance as the inspector reflects the programme's broader movement away from hierarchy, juxtaposition and tension towards a more assured, settled status quo. Marshall's relaxed, self-effacing demeanour plays a crucial role in structuring Humphrey's character.

Crucially, and contra prevailing trends, *Death in Paradise* completes the transition by confidently embracing conventional episodic form and rejecting serial elements. The protracted romantic tension between Humphrey and Camille, which continues that between Richard and Camille, disappears with Camille's departure from the programme in series four. Although Camille is replaced by another female character, DS Florence Cassell (Joséphine Jobert), the series avoids the easy trick of substituting Florence for Camille as Humphrey's 'love interest'. In place of romantic frisson, appropriately, Florence and Humphrey develop a friendship – one of several at the heart of this series. *Death in Paradise* in the Marshall era thus moves away from contrast as a route to comedy towards subtler, more everyday, 'episodic' pleasures of familiarity and friendship, via a wholehearted embrace of episodic form.

Appreciating the Episodic Performance

In the context of prevailing attentiveness to the achievements of various complex television serials, this chapter took up an alternative focus: the episodic series, often neglected by aestheticians and cognitivists, and undervalued by critics. Yet the analysis herein might nevertheless pertain helpfully to studies of long-form serials. Although episodic and serial forms have historically been juxtaposed, and can reasonably be distinguished at a conceptual level, the two are inevitably interdependent. The presence and development of serial elements within episodic programmes is frequently noted, but the importance of episodic qualities within serial form, in comparison, is more often overlooked. When considering serials, it must be remembered that episodic form remains fundamental to the art of television. Even the longest-running, most complex serial depends necessarily upon episodic structure, its individual episodes exhibiting discrete elements to a lesser or greater extent; occasional episodes exhibit conspicuous singularity, their episodic qualities heightened and celebrated ('one-off' specials and bottle episodes, for example). Seriality and episodic form are inextricably connected; the very concept of a serial (or a series) depends upon the existence of an episode, and vice versa. Therefore to appreciate fully the long-form serial, it is necessary to appreciate the episodic.

The episodic series also warrants dedicated attention in its own right, especially in relation to performance. Within necessarily limited space, I have gestured towards some of the specific implications of episodic structure for performance, and have advocated a reappraisal and revaluing of this form. In celebrating the navigations, negotiations and achievements of one specific series, *Death in Paradise*, I hope to have shown that it is possible to proffer sensitive and appreciative explorations of the particularities of performance within traditional episodic series. Performances in an episodic context offer both the pleasures of nuanced repetition and familiarity, and those of immediacy: we value and may evaluate these performances for the ways in which they augment our existing perceptions of each character, and we also relish them for their own sake, for the delight they elicit in the transient moment, regardless of their contribution or otherwise to more expansive narrative arcs. Admiration for works in long-running serial form is not incompatible with appreciation for the artistry involved in creating simple, 'escapist', dramatic episodic television fiction. Sometimes good things come in small packages.

Notes

1   Specifically, this chapter adopts the mode of interrogation most often associated with television aesthetics: detailed stylistic analysis and evaluative criticism of a particular programme. I aim to offer a sustained, appreciative account of the programme's singular achievements as they relate to this collection's core focus: television performance. For a concise introductory overview of 'television aesthetics', its principles, concerns and methods, see Cardwell (2006).

2   See, for instance, George Toles' account of Don Draper in *Mad Men*, in Jacobs and Peacock (2013: 147–73) and Elliott Logan's exploration of Walter White in *Breaking Bad* (2016).

3   Episodic programmes, especially middlebrow dramatic or comedic series, often fall under the umbrella of 'invisible television': popular television that is nevertheless mostly ignored by TV scholars; this phenomenon is addressed in a special issue of *Critical Studies in Television* (2010), 5.1.

4   Blanchet and Bruun Vaage use the term 'series' indiscriminately, to refer to (episodic) series and/or serials. All their detailed examples are serials.

5   Reciprocally, undue emphasis on seriality and the concomitant neglect of episodic form also obscure the salience of episodic structure – and episodic singularity – within long-form serials. Seriality and episodic form are inextricably connected; the very concept of a serial (or a series) depends upon the existence of an episode, and vice versa.

6   'Character development' can refer also to the actor's preparation – their work before and during performance, something which is clearly pertinent to both serials and episodic series. Herein, I refer rather to 'diegetic' character development: that which is discernible within a work, 'written in' or 'drawn out' via performance across time. It is this type of character development that is more strongly associated with serial form.

7   It is interesting to note, however, that some cognitive studies show we often like most those whom we do *not* come to know as deeply: 'Although people believe that knowing leads to liking, knowing more means liking less' (Norton *et al*. 2007: 105).

8   Perhaps the correspondence between engagement with episodic series and with people in real life is overlooked by cognitivists because, despite their focus on characters, they too rarely mention *performance*, which is after all the initial means by which we are engaged.

9   Occasionally, a case runs across two episodes.

10 The condemnation of *Death in Paradise* as 'escapist' television is closely connected with its conservative episodic form. Cognitivists have revisited 'escapism', arguing that it is not something to be embarrassed about, for it merely exploits a fundamental human tendency to avoid 'cognitive taxation' and allows us to 'exercise our rights as cognitive misers' (Raney 2004: 362, 364).

11 *Death in Paradise* exploits Miller's background in comedy; in doing so, it continues a convention of casting comic actors in light-hearted episodic series (see, for instance, Martin Clunes in *Doc Martin* (ITV, 2004–) or Alan Davies in *Jonathan Creek* (BBC, 1997–2016)). Miller's successors in *Death in Paradise*, replacing him as the principal inspector, continue the trend: Kris Marshall, from 2014, and Ardal O'Hanlan, from 2017, best known respectively for their roles in sitcoms *My Family* (BBC, 2000–11) and *Father Ted* (Channel 4, 1995–98).

12 Miller's performance as Poole undoubtedly exploits stereotypical notions of Englishness, his demeanour contrasting broadly and expressively with the looser, more relaxed island milieu. However, the particular characters and performances of his Saint Marie colleagues are carefully individuated, as previously mentioned: the core team comprises insouciant Dwayne; eager and assiduous Fidel; sharp-witted, often impatient Camille; and the smooth-talking, careerist commissioner. Indeed, the programme mostly eschews simplistic stereotypes and caricatures of nationality and race, both in terms of its central characters and cameo roles.

13 *Death in Paradise* appears reluctant to abandon its original premise when recasting, sticking with its format of a British (or UK) inspector adapting to life on an exotic (though by this point, for regular viewers, simultaneously comfortingly familiar) island.

This may be the reason that the writers appear to have overlooked the dramatic possibilities of promoting the current second-in-command, and occasional action hero, DS Florence Cassell (Joséphine Jobert) to the central role. However, while the programme's initial concept might justifiably continue to determine the (non-native) nationality of the inspector, it does not explain why all three actors to have held that role have been white men; this obduracy contrasts with *Death in Paradise*'s otherwise balanced and inclusive casting. Another study might fruitfully explore the show's representations of race and gender; obviously that falls outside my remit here.

14 Kris Marshall's entrance here, his performance and the aspects of his character thereby established clearly echo his previously best-known role in *My Family*. His inspector builds upon the expectations of viewers familiar with that earlier work.

2

The Performing Lives of Things: Animals, Puppets, Models and Effects

Jonathan Bignell

In the first episode of the adventure series *The Prisoner* (ITV, 1967–68), the unnamed protagonist (Patrick McGoohan) finds himself a vantage point so he can look out over the village to which he has been abducted. A long shot from his point of view takes in the main square, whose ornate Mediterranean-style buildings frame a grassy lawn with a central fountain. Colourfully dressed people stroll around it, but suddenly the villagers are instructed to freeze by an authoritative male voice over a public address system. They all stand stock still apart from one man who runs frantically across the square. A small white bubble or ball appears on top of the water spouting from the baroque stone fountain, inflating rapidly until it is about two metres in diameter. The Prisoner looks on as the ball leaves the fountain and, seemingly of its own volition, chases the fleeing man, skimming along across the grass, halting him and appearing to incapacitate him by covering his face so that he cannot breathe. This is one of the first sinister and puzzling events in The Village, where the protagonist, known only as Number 6, is repeatedly interrogated and his attempts to escape are continually frustrated.

The white ball, known as Rover, appears in nearly every episode of the series and makes a similar kind of pursuit in a later episode, 'Checkmate', but mainly it acts as a physical obstacle that blocks a person's path or forces them to move in a certain direction; it is a threat more than an active aggressor. It behaves as if it can respond interactively to its victims, anticipating their actions so as to prevent or modify their behaviour. It is aware of its own spatial position, the positions of others, and the possibilities for action presented by the local environment. It polices and monitors the prisoners in

The Village, like the guards in a wartime internment camp, and like a guard dog or sheepdog it can harass, chase, block or attack its victims. It is occasionally glimpsed sitting passively in the ball-shaped Aarnio chair usually occupied by Number 2, who runs The Village from his underground control room. Occasionally it sits in the corner of a room, or glides without apparent purpose around The Village. Like a gun dog, it can retrieve the victims of its attacks, appearing to transport their unconscious bodies from where they have been incapacitated, in the adjoining sea or on the wooded perimeter, back to The Village. In the episode 'Free for All', three small Rovers work in a group to transport the Prisoner across the sea to The Village after he is recaptured. In the same episode, Rover is encircled by four men, seated with their arms folded, each gazing at its luminescent surface in an underground chamber cut out of the rock beneath The Village. Here, Rover is something like a television set, or perhaps some kind of meditation device with which the men have a mental connection. Across the series, it functions like a human character, like an animal, and like an object or machine (Figure 2.1).

Figure 2.1   Rover and Number 6 (*The Prisoner*, ITV, 1967–68)

Rover is not a character in a conventional sense, but this chapter addresses how critical analysis might describe and evaluate such non-human 'performances' in television fiction, and how they affect distinctions between actor and role, and between character and narrative function. Across the history and genres of television, there have been very many 'objects' that narratives make expressive but that are not human, nor even, in some cases, alive at all. The white balloon-like Rovers of *The Prisoner*, if they are to play their part in the fictional world, need to function as expressive 'performers'. In a way that incidental props or scenic components do not, such non-human performers need to seem like subjects as well as objects, so that they can play a part in storytelling rather than forming part of the backdrop against which action happens. So, the chapter has questions in common with those addressed by the discipline of Animal Studies, which includes investigation of how creatures that have hitherto been treated as objects by dominant human subjectivity might be thought of as subjects too (Derrida 2002). If those creatures that were formerly other become comparable with or even equivalent to humans, the boundaries between self and other, human and non-human, are destabilised (Haraway 1991, 2003). People become more like things, and things acquire some of the attributes of people; theoretical work on pets and artificial creatures (such as cyborgs) has debated the recasting of unexamined cultural and political hierarchies that this reconsideration produces. This chapter develops some of these ideas in relation to performance by asking how animals and objects behave in programmes as if they were characters, and how this affects the building of fictional worlds in those programmes.

It is customary to assume that one of the qualifications for action to become a performance is that there is an intention to express, whether by an actor performing a role or a non-actor in an unscripted self-presentation on a game show or in a documentary. Objects and animals problematise this assumption of intention, and the chapter focuses on how the relationship between viewer and non-human performance renders boundaries between self and other, or alive and inert, fluid. For this chapter, the key issue is not intention but the embedding of animals within mise-en-scène, troubling the distinction between the living and non-living components of the fictional world. The chapter returns to inanimate objects later, to consider puppets and models that are integrated into created settings and combined with human voice-over performances in the action series *Thunderbirds* (ATV, 1965–66). The series has an explicit focus on action, vehicles and special effects and it is the significance of integration into narration and setting, the work expended to fit 'things' into mise-en-scène, that the chapter highlights. The 'performance' of animals presents some of the same problems of intention, expressivity and embedding, and in keeping with the 1960s period of the other examples the chapter addresses the Australian filmed series *Skippy The Bush Kangaroo* (Nine Network, 1968–70), whose modes of performance followed from earlier adventure series featuring expressive animals, such as *Lassie* (CBS, 1954–73) and *Flipper* (NBC, 1964–67).

Semioticians interested in animals' effective use of systems of signs sometimes reject the distinction between natural behaviour and staged performance (such as Paul Bouissac's work on circus (1981), showing that animals repeat behaviours natural to their species within a location marked out by humans for performance). Skippy the kangaroo was portrayed as an intelligent, helpful and usually compliant companion to nine-year-old Sonny Hammond (Garry Pankhurst), living with his father Matt (Ed Devereaux), the Head Ranger at Waratah National Park Wildlife Reserve, north of Sydney, Australia. Storylines in *Skippy* concerned rescues, encounters with visitors to the park, protecting the park from rustlers, smugglers and escaped prisoners, and the domestic tensions between Sonny, his father, and Sonny's older brother Mark (Ken James). In the pilot episode, 'Man from Space', for example, Sonny sends Skippy to fetch help when they discover a downed pilot in the park whose parachute is stuck in a tree. Skippy exhibits the capacity to move, think and communicate, and to engage in planned sequences of action, pretence or play, for example. She is at an intersection between a human world and the wildness of the 'other'. Like a human character, the camera attributes her with a point of view, and she sometimes becomes a subject who can look at and look back at other performers, as well as being an object of the camera's and other characters' looks. She is a companion and helper for humans in the series, with agency and character, but always subject to narratives driven by people rather than herself (Figure 2.2).

Skippy's role is somewhat like that of a family pet, living with the Hammonds, their workmates, and the pseudo-familial female characters of teenage house-guest Clarissa 'Clancy' Merrick (Liza Goddard) and potential love-interest research scientist Dr Anna Steiner (Elke Neidhardt). But Skippy is intelligent; her actions protect the family and the park against threats from outsiders such as Dr Alexander Stark (Frank Thring), the owner of a private zoo who attempts to kidnap her. She can not only understand instructions,

Figure 2.2 Skippy's night-time rescue of Sonny (*Skippy The Bush Kangaroo*, Nine Network, 1968–70)

and communicate by means of gesture and vocal sounds, but also in later episodes she is able to open doors and operate the buttons and knobs of the radio in the Ranger Station (to communicate with Ranger helicopter pilot Jerry King (Tony Bonner), for example), tie and untie knots to facilitate outdoor rescues, and play the drums in a comic musical sequence. Skippy's role was performed by moving around in outside (and sometimes interior) locations; adopting static poses for shot reverse-shot exchanges and implied point-of-view shots; and details of action in close-ups involving her arms and claws. In other words, the conventional rhetoric of continuity editing and the hierarchy of long, medium and close shots in a coherent filmic space were used to suture the kangaroo into the narrative like a human character.

Thunderbirds is set in 2064 and concerns the activities of the secret International Rescue organisation, operating from a Pacific island. In response to natural disasters and failures of technology the Thunderbirds rescue vehicles, piloted by the sons of former astronaut Jeff Tracy, perform extraordinary feats in the air, in space, underwater and by tunnelling underground to rescue survivors. The human characters are 'performed' by puppets, in highly detailed sets, and the advanced aircraft, monorail trains, mining machines and space satellites featuring in the storylines are represented by models. The puppets of *Thunderbirds* are also shot following the conventions of continuity editing, and human speech and familiar sounds are used to stitch together a coherent fictional world. The conventions of vocal modulation for relative distance from the camera's point of view, for example, are adapted into the model environments. Bodily and environmental sounds correspond with physical action, such as footsteps matching puppet movement, and characters emit cries when falling or being shot, laugh, or grunt with exertion. The futuristic vehicles make sounds deriving from library recordings of 1960s cars or jet aircraft, for example, and there are numerous sound effects for models representing hydraulic ramps and electrically powered elevators, and the frequent explosions in disaster sequences use stock sound effects of dynamite blasting. This sonic material produces an impression of veracity, yet it potentially emphasises how familiar sounds have been contained, reprocessed and adapted for a new context. The fictional world is co-terminous with the real, because there is resemblance but also distinction between them, and each fiction is at one level self-reflexive in its performance of verisimilitude and world-building.

The models and puppets in *Thunderbirds*, representing people living with supremely capable machines embedded in a flawed technological utopia, express and explore relationships with commodities and the lives of the things that humankind makes. The life of things was an aspect of Marx's analysis of the commodity in *Capital* (1954) where he argues that in capitalism the things made by human labour seem alive, inasmuch as they have social meaning and mediate relationships between people. Inanimate things take on a life that derives from the accumulated human labour that made them, but which is hidden from view. People and their labour, by contrast, are occluded, perceived in terms of quantitative measures like labour costs and productive value, and seem like things. This chapter aims to show that things on screen work over the problem of the otherness of the non-human, and what such a relationship with the other hides and reveals. Flipper is not simply a dolphin, and Skippy is not just a kangaroo; rather, each is represented as a part of the natural world that is sufficiently self-aware and intelligent that it can take part in television storytelling, though the stories are always about a human world. Similarly, in *The Prisoner*, Rover is not simply a white ball, but enacts the power of The Village's controllers over its inhabitants and is a means for the narrative to express that power concretely.

The series themselves are commodities, made for a transnational television market. *Skippy*'s creator Lee Robinson designed the format for sale to the USA as well as the relatively small domestic Australian market, in which imported US drama predominated. On a research visit to the USA he was inspired by *Flipper*, in which a widowed park ranger and his two sons befriend a dolphin (Anon. n.d.). Placing a kangaroo at the centre of Robinson's format, prominently featured in the series' opening credits, signified Australian distinctiveness alongside cultural values of outdoor vigour and male sociality ('mateship'), announced in the whimsical but jaunty banjo theme song naming Skippy as 'our friend ever true'. Australian TV mogul Frank Packer financed the series, which sold to 128 countries. Both *The Prisoner* and *Thunderbirds* were made with the backing of the television mogul Lew Grade, who ran the British ITV company Associated Television (ATV) and whose Incorporated Television Company (ITC) was at the forefront of programme export to the USA (Bignell 2005). Grade produced *Danger Man* (ITV, 1960–68) in which McGoohan starred as a secret agent for NATO, and having sold the series internationally Grade backed McGoohan's concept for *The Prisoner*, which could be regarded as a follow-up with the same protagonist. The creators of *Thunderbirds*, Gerry and Sylvia Anderson, had been making puppet series in the popular genres of the western and science fiction as a low-budget way to enter live-action television production. But the lucrative merchandising of toys and licensed products from their puppet action series *Supercar* (ATV, 1960–61) and the sale of *Fireball XL5* (ATV, 1962) to the US NBC network led to the creation of *Thunderbirds*, aimed at the international market. The focus of this chapter is on the 'things' that appear in programmes, but it is also germane that storytelling about those things was a successful economic strategy to create value by disseminating television programme commodities for international trade.

Framing Performance

The representational illusion of screen fiction, in claiming to offer the presence of a character on-screen, is potentially fractured by the awareness of performance as the medium through which this representation has been created.

But the predominant illusionism of television fiction makes representations work hard to seem like presentation, occluding the work done to create an illusion (Heath 1981: 113–30). It follows that acting, as one of the crafts comprising filmmaking, should conceal conscious effort or self-consciousness. The problem with human actors is that they are self-aware, Lev Kuleshov (1974: 99) proclaimed, arguing that natural movement on film could be best exemplified by the deft and economical motion of a skilled factory worker, or 'the filming of children or animal movement' because of its 'profound innocence, naturalness, and simplicity'. The point is to eschew exaggerated theatrical posturing, so that body, setting, and perspectival space in general could be an energised, organic unity oriented for the spectator, organised by a master point of view as established by Renaissance perspective. Whether naturalness is achieved by extreme self-consciousness (using and controlling the body as object or instrument), or extreme consciousness (embodying, being or becoming the character), the aim in conventional screen acting is to merge actor with role. Mechanical, animal or inhuman action most appropriately serves the ideologies of filmic representation that are their master, implying an unmarked power dynamic that underlies the effectiveness of the ideal cinema that Kuleshov describes.

The disavowal of the self-consciousness and conventionalisation that performance requires was emblematised for Kuleshov by animals, children or workers wholly absorbed in a task. Such absorption minimises the distinction between the performance and activity before and after it when the performer is not in role. Jean-Louis Comolli (1978) makes the distinction between the body of the actor in a historical film and the person that the actor is playing. He argues that there will always to some degree be a split in the spectator's belief, between accepting that the body on-screen is the character being portrayed and the awareness that there is an actor with a life outside the role. There is 'a body too much', and the two bodies are in tension with each other, producing a potentially critical space for the spectator's awareness of the work of cinematic signification (Bingham 2010). Moreover, nuances of performance result from the detailed choices made by actors, because of their specific life histories, professional training regimes and bodily characteristics and capabilities (de Cordova 1991: 119). These particularities are exposed by the 'commutation test' (Thompson 1978) that analyses a performance by imagining how a role would be different if adopted by a different actor. For animal and puppet performances, there can be no commutation because the things in the role are either unique or replaceable by an almost identical copy. It does not matter how many weather balloons were used in *The Prisoner* to embody Rover, nor how many kangaroos were used in the filming of *Skippy*, because to most viewers the balloons and kangaroos are indistinguishable one from another. It does not matter which of the differently scaled puppets of the Tracy family is used in particular shots in *Thunderbirds* because they are made to be exchangeable with one another. This seems to suggest that animals and things cannot perform as actors, because they have no unique identity to bring to the role they play. In this sense, they cannot be performers, but in another important sense they fulfil the ambitions that screen acting theorists like Kuleshov have had for a century. Animals and puppets become embedded seamlessly in the mise-en-scène, taking part in performance in the sense that performance means the total signifying world within the frame.

Rover belongs in the world of *The Prisoner*, inasmuch as some of its characteristics align aesthetically and functionally with other aspects of The Village. As Mark Bould (2005) points out, Rover matches the circles and spheres in *The Prisoner*'s production design, seen in the Village emblem of the penny-farthing bicycle, the circular control room, the Aarnio ball chair, spherical camera mountings and aerial shots of unfurled umbrellas. Rover is something like a beach ball, and The Village seems designed to resemble a seaside holiday camp in which the inhabitants live in chalets, take part in organised leisure activities, adopt a colourful and casual slacks-and-blazer dress code, are addressed by a loudspeaker system and may use canopied bicycles. All of these features of The Village were true of Butlins and Pontins holiday camps in 1967, for example the one at Pwllheli a few miles from *The Prisoner*'s Portmeirion location. The Village might be run by the British spy organisation to which we assume Number 6 belonged before his abduction, and connotations of Englishness include college scarves, seaside deckchairs and the clipped upper-class tones of the public address system's announcements. All these details might be a ruse in order to manipulate captives, however, with The Village an internment camp like the one at Inverlair Lodge in Scotland that suggested the *Prisoner* format to its creators, McGoohan and George Markstein. Certainly, Portmeirion, a coastal estate in Wales built by the eccentric architect Clough Williams-Ellis in the 1920s, confuses geography by whimsically blending buildings that pastiche architectural styles from different periods and regions of Mediterranean Europe. Moreover, interiors in *The Prisoner*, shot in a studio, have modern, fashionable and technologically advanced furnishings, and the Village control room is distinguished by its metallic polished surfaces, large projected images and surveillance equipment, next to machines for mind control and remote manipulation. The confusing spatiality of The Village and the hi-tech apparatus in it alongside a whimsical and often traditional appearance, together with the repeated failure of the fictional world to conform to the expectations of consistency and plausibility, place the viewer in the same situation as Number 6 himself. He is repeatedly frustrated in his attempts to find out geographically and existentially where he is and what his experiences mean, producing a hesitation (Todorov 1975) about whether the drama should be framed within the conventions of fantasy, reality or allegory. Rover takes part in producing this uncertainty, for it exists and acts in the fictional world in uncanny ways and its nature is never explained. Like The Village as a whole, it is a given fact and also an enigma suggesting that there is a greater but unseen significance to which it points.

The design aesthetic of *Thunderbirds'* fictional world combines the influences of twentieth-century aircraft design and modernist architecture, as does much television science fiction of the era (Britton 2009: 342). The Thunderbirds rescue craft have streamlined, smooth surfaces and swept-back wings designed for high-speed flight, and the space rocket Thunderbird 3 drew on visual references to the clustered boosters of the Russian Soyuz spacecraft of the 1960s, for example. But the vehicles' appearance also departed from functionalism in order to portray character, most obviously for Thunderbird 2 whose stubby wings signify its support role in heavy lifting and cargo carrying. The architecture of the Tracy Island complex from where International Rescue operates, and other buildings such as airports, military bases and hotels in the series, very closely matches the unornamented, geometric steel, glass and concrete construction of twentieth-century buildings. On the other hand, the Palladian style of the mansion inhabited by Lady Penelope Creighton-Ward, International Rescue's London agent, was based closely on eighteenth-century Stourhead House in Wiltshire in order to signify upper-class Englishness. The design palette used in *Thunderbirds* emphasises the strong colours of blue, red, yellow and green, and the silvery polished metal alloys from which its vehicles are supposedly made. But although the series primarily has a techno-utopian aesthetic, its makers were careful to add realistic detail such as dust and tyre marks on roads and runways, and smoke stains on engine exhaust manifolds. The premise of *Thunderbirds* is that, as in real life, machines are subject to breakdowns, malfunctions, accidents and sabotage that cause them to fail and endanger humanity, but the Tracy family group of more capable humans, with their more capable machines, preserve a fragile international world order based on mid-twentieth-century ideologies projected into its futuristic setting (Bignell 2011).

Skippy belongs in the landscape of Waratah National Park in a way that human actors do not; like Rover in *The Prisoner*, the kangaroos in *Skippy* are part of the represented fictional world, appearing natural and embedded in it, in contrast to the human performers. In *Thunderbirds*, the human characters of the stories are represented by puppets who mediate between living bodies and constructed objects. Indeed, close-up sequences in *Thunderbirds* feature human hands operating telephones or opening drawers, for example, suggesting the unmarked equivalence of the human body and the puppet body. In his classic study of film acting, James Naremore (1988) distinguishes between presentational and representational kinds of performance. Presentational performance, as in comedy or the musical, shows an awareness of the performance being directed towards an audience, and sometimes even an address directly to the audience present in the cinema (Brown 2012). By contrast, the majority of narrative films are representational, in the sense that there is no acknowledgement of an audience and performance is naturalised as part of the fictional world. Animals are both presentational and representational performers simultaneously. The same can be said of objects like Rover, which show off what 'things' can do but at the same time are part of the world in which they exist because they are not aware of what they are as objects, characters or performers. Acting might seem to be a secondary mode of being that is built on top of an authentic self, the actor, whose body and voice are used to create an alternate fictional identity. But, as Philip Auslander (1997) argues, the notion of the real actor is projected backwards as an anterior reality created by the performance that it supposedly underpins. In works using non-human figures, an anterior, enfolding reality is over-signified in order to create the impression of the real, by means of detail of setting, sound and voice.

Theories of performance do not always assume intention (Carlson 1996: 39–41). As long ago as Erving Goffman's *The Presentation of Self in Everyday Life* (1959) work on performance has stressed the audience rather than the actor, defining performance as activity that occurs over a specified period of time, during which there are observers who witness the action. The actor does not need to be conscious of his or her role, and the important factor is a relationship between performer and audience that enables the observers to recognise what they see according to a particular set of social rules about behaviour and its contexts. The audience's knowledge of these rules, and their invocation of them, creates performance. It is not necessary to assume that the actor is presenting deliberately to a spectator, nor adopting a specific role of which he or she is conscious. For Goffman, it is the act and process of staging for someone, of framing something (on stage, in a framed image) that defines performance. It is this approach to performance that this chapter adopts, because it stresses that performance is a kind of making, shaping and showing that foregrounds mise-en-scène.

The Work of Mise-en-scène

The notion of mise-en-scène, combining the meanings of putting on stage, placing into a setting or directing a show, draws attention to the fact that television programmes are a work of labour aimed towards an audience; they are a showing for someone. To be interesting and dramatic, animals and machines need to be embedded as part of a naturalised world in which they are believable, even ordinary. But on the other hand they are protagonists or antagonists, and as such they are motors of narrative rather than merely aspects of setting, unlike the other contingent objects or props that surround them as the given circumstances of their 'lives'. So, attention is always drawn to them. Because of the limited degree to which animals and machines can be expressive, there is excessive expression in the television programmes in other ways. The actions of animals and machines are enfolded with music, sound effects, point-of-view and shot reverse-shot patterns that provide narrative tension, pace and dramatic shape. Storylines are designed to offer sequences in which the animals and machines can show off their capabilities and perform against the backgrounds of realist settings surrounding them, but this also produces a tension between non-human performance and the human. In representational performance, there is a tension between the body of the actor and the belief in the presence of the character on-screen; what Comolli (1978) termed 'body too much,' as discussed above. Naturalistic storytelling reduces this as much as possible, but the outcome might be that the technique of performance is foregrounded as much in its concealment as in its overt presence. This is the contradiction at the heart of method acting, in which the ability to merge with the role is precisely an attribute associated with star actors (like Dustin Hoffman or Robert de Niro, for example). The technique and provenance of the performance become dominant. With animals and machines, there is no story of human training or experience to form this provenance. But instead there is craft and preparation. The labour of production and the skill of its achievement are on show and at the same time are occluded by illusionistic representation.

The original design for Rover in pre-production was that it would be an autonomous vehicle or robot about half the height of an adult, running on wheels (Pixley 1988: 11). It would have been circular, with an inflatable skirt like a hovercraft, with a black-and-white domed cupola top. This specially built prop sank during the filming of test sequences in the sea off Portmerion, and was replaced by a white latex weather balloon controlled by long, thin wires that allowed it to be dragged across land or water. Its movement could sometimes be made very rapid or its wobbling oscillations uncanny by speeding up film or running it backwards. Its thin, flexible skin allowed internal luminescence to be seen, achieved by back-lighting, and post-produced roaring sounds were added for shots of it moving. Each of these features adds to the impression that it is alive and autonomous, able to move equally quickly over land or water, or through the air, over long distances and at a speed faster than a running man. Its relationship with the controllers of The Village is clarified when they declare an 'Orange Alert', when the formation and return of the Rover is projected on a giant screen in the Village control room. The screen shows a repeated stock shot of an underwater location, presumably representing the sea near The Village, from where Rover inflates from the sea floor, detaching itself and floating quickly upwards. In the episode 'Free for All', Rover returns to this underwater location (the inflation sequence runs in reverse), and in the final episode, 'Fall Out', it moves down into a hole or tunnel into a cave where it deflates to a smaller size and stays at rest. Making Rover's performances seem plausible took considerable work, which is concealed when the finished sequences appear in the series, but on which the fore-grounding of The Village's distinctive uncanniness depends. Rover's apparent lightness, plasticity and insubstantial form, potentially signifying its vulnerability, become all the more menacing and uncanny when hidden labour succeeds in giving it the apparent ability to move independently, initiate and respond to situations strategically, and enact violence on Village inhabitants.

The world on-screen is a performative space, whereas the space of production is a working, craft space. The aim of that craft is to create performance in a way that is believable, by screening off, out of frame, the labour of production and the time before and after the performance. For *Skippy*, New South Wales Minister for Lands Tom Lewis, heading the newly established National Parks & Wildlife Service, gave permission to film and to build standing sets in Ku-ring-gai Chase National Park, north of Sydney. The Hammonds' house, with associated roads, helipad, power and water supplies, was built amid 24 acres of land, with access to a further 500 acres. Animal trainer Scotty Denholm used nine female kangaroos to portray Skippy, keeping each one dormant in a sack until needed because of their tendency to run off into the bush. As well as shooting the kangaroos static and moving on location, a stuffed kangaroo was used for rear shots, and kangaroo forearms (sold as souvenir bottle openers) were manipulated by the crew in close-up shots requiring specific, dextrous actions (BBC, 2010). The crew gave the kangaroos chocolate, gum or grass to chew, and occasionally wrapped an elastic band around the animal's lower jaw, so that Skippy would appear to vocalise. Skippy's communication by making clicking sounds was a post-produced effect, and dramatic music stings and music matching physical action ('mickey-mousing') pointed out turns and moods in the narrative. It is the tension between foreground and background environment, and between the non-human characters' ability to act as well as to be acted upon, that makes these programmes pleasurable and potentially troubling. It also makes them potential resources for critique of conventional expectations about living presence, performance and the body, and the ethical relationships of human to non-human bodies and the 'things' in our environment.

Thunderbirds was made in elaborate soundstages, because the emphasis on colour and visual detail in the mise-en-scène required shooting on high-speed 35 mm film and high-powered lighting. There was one soundstage for models and special effects, and two for puppetry (Richardson 1991: 6). A gantry for up to seven puppeteers was built to a height of nearly three metres above the floor so that the puppeteers did not have to stand next to a puppet to operate it, and thus much larger sets could be built. The need for vehicles at (usually three) different scales, for shooting in long, medium and close-up shots, meant considerable workload for the model-makers, and for the puppets portraying the 13 regular characters there would be heads with three different expressions so that they could be changed during shooting, and a backup set in case of damage or malfunction (La Rivière 2009: 110). Shallow water tanks were built for the establishing shots of *Thunderbirds'* Tracy Island and seaborne rescues, as well as flat rostra surrounded by cycloramas of sky for land-based action. As early as 1957, the company had developed video assist so that the production crew could see what the film camera was recording (Sellers 2006: 80); invented electronic lip-synch to match a pre-recorded voice track to the solenoids activating the puppets' mouths; and for *Thunderbirds* a rolling road was created for sequences of vehicle movement. Three different rolling surfaces could be used, in the extreme foreground, middle ground and background. When moving at different speeds (the closest to the camera moving fastest) and shot from a low angle, a strong sense of depth perspective is created, like a deep focus tracking shot in a live-action film. While the visual grammar and soundscapes of *Thunderbirds* aimed to conform as closely as possible to existing convention, this required significant technical innovation (Figure 2.3).

Television as a technology of representation and as a cultural form (Williams 1974), with its attendant ideologies and conventions, is a condition of, and conditions, how performance on screen works. The performance of objects or animals in fictional drama depends on finding or constructing something that will then be put on show. This kind of showing changes what is put before the audience, producing an expectation of significance over and above what it meant in ordinary social reality. Issues of context and the delimited location of performance activity are explored in Goffman's *Frame Analysis* (1974), in which framing draws on Gregory Bateson's theories of play and fantasy (1955). Both play and performance are framed by rules that allow them to be non-referential, and the actions they comprise can be potentially socially unacceptable in ordinary life. The numerous violent acts, destructive events and interpersonal cruelties of television fiction are enabled

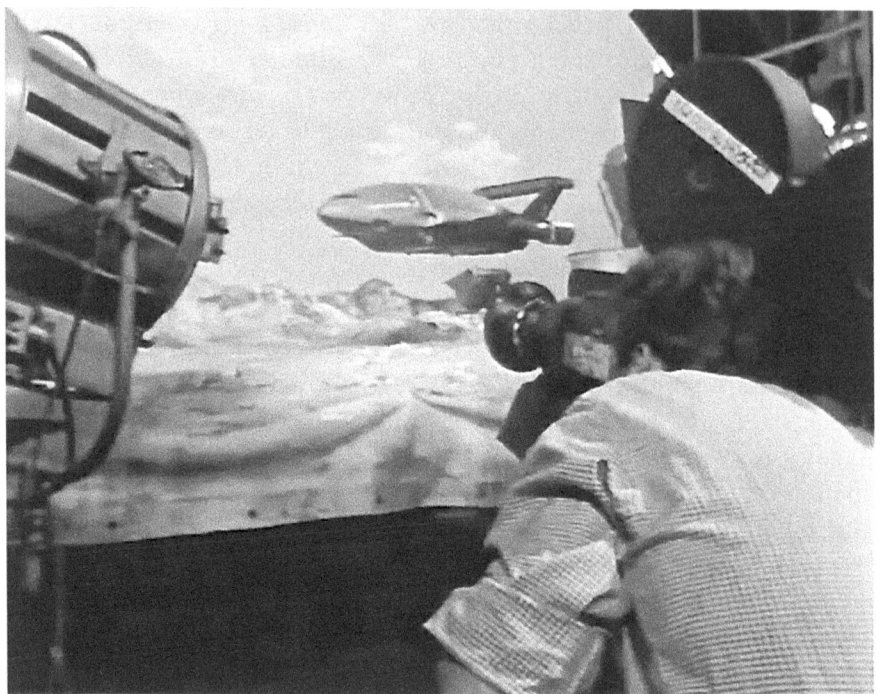

Figure 2.3 Filming Thunderbird 2 (*Thunderbirds*, ATV, 1965–66)

and constrained by the modalities of play and fantasy whose framing makes them acceptable to their audience. Moreover, the metaphor of framing links nicely to the physical frame of the television screen, which demarcates a spatial boundary in which representation, fiction, performance and relayed actuality are made possible subject to the codes and conventions of the medium.

As an analytical concept, framing brackets off the question of intention in order to focus on the relationship between the audience and the performance, as a relationship in which meaning is actively made by the spectator. It alludes to the compositional frame of the television image and the processing of visual and aural information geared towards the production of fictional, entertaining and engaging storytelling within the conventions of television formats and genres. Furthermore, the concept of framing demarcates the specific professional, institutional and technological activity of making programmes in a medium embedded within the cultural forms of a specific society. Goffman's version of framing is similar to the concept of ostension developed by Umberto Eco (1977) which addresses everyday behaviour in light of comparisons with theatre. Adopting a version of semiotics closer to Charles Pierce's notion of language use than Ferdinand de Saussure's approach to language as system, Eco focuses on how performance places something on view for an audience to perceive. Ostension is a kind of showing, in which something is selected, and treated by processes of dressing and arrangement so that an audience will be able to decode its meanings. Like framing, ostension draws attention to the spatial and temporal specificity of a performance event, and how it is set within particular social and cultural codes. The ostended object or animal is not only there in itself, but invites the audience to consider how it may link outwards beyond itself to a greater representative, exemplary, symbolic or metaphorical significance. For *The Prisoner*, Rover emblematises the perverse but purposive constraint of individual action by the institution of The Village, which itself has connotations of Cold War political totalitarianism and existential anxiety. Skippy the kangaroo is tame but wild, a character and an instance of inhuman Nature, and an icon of the many meanings of Australian-ness. The puppets who play the human Tracys in *Thunderbirds* are objects like the machines they use and the settings in which they 'live', but the role of those humans is to master machines and futuristic environments in order to rescue other humans from the threats they pose.

The mise-en-scène of the programmes I have referred to is a precondition for their non-human (and human) performances. Strong lighting, bright colours and voice, sound and music matched to action show off the detailed materiality of the performing 'things', and their settings and background effects. The performances are set within spaces that are shaped to make the most of meticulously crafted models and professionally trained and managed animals. Each of the programmes discussed here uses extensive 'practical', functioning effects: sequences shot as-live in front of the camera (not post-produced electronic effects) that equally require the expertise of a team of specialist workers. In addition to the performances of the animals and the objects on-screen, these programmes display the skills of those workers to the audience. The performances of animals and objects work together with visual design, sound effects and music to establish their distinctiveness as television fiction, but within the broad generic conventions of live-action adventure drama. There is a politics of this kind of performance, but it is often unremarked, and this chapter has begun to address the ways in which such a focus might be developed. Attention to the detail of mise-en-scène in these programmes can show the potential troubling of the boundaries between subjects and objects of the camera's gaze, and between human and non-human others, that is largely contained by strategies of storytelling and world-building that aim to master it. The performing 'lives' of 'things' work with and through questions of aesthetics and representation that are central to the forms of work, value, organisation and articulation in television fiction.

3

The Enduring Act: Performance and Achievement in Long Television

James Walters

Introduction

This chapter seeks to evaluate the place and weight of performance in long-running television programmes. By focusing on a small selection of key examples, I hope to illustrate some relationships that I take to exist between scale and depth, and the achievements of certain actors as they handle these interactions. I choose the framing descriptive term 'enduring' in part to reference the longevity of particular screen performances that I offer for consideration but also to underline the *work* of actors as they commit to this wide span of screen time: their acts of endurance. This interest seems, if nothing else, appropriately timed. One of the hallmarks of the so-called new 'golden age' of television (Lawson 2013; Leslie 2017) is the widespread establishment of programmes that commit themselves to numerous seasons and, as a related consequence, develop involved plots that stretch across many accumulated hours of television. As Adrian Martin suggests:

> The great strength of the recent crop of episodic TV production has been in its development and exploration of 'long form narrative' – stretching, sustaining, and exploring the byways of a story and its complex world. This gives rise to a particular type of in-depth involvement on the part of spectators.
>
> (Hilmes et al. 2014: 26)

Indeed, for the dedicated viewer, persevering with such programmes can offer enticing rewards as patterns and motifs emerge and repeat that privilege the close scrutiny enjoyed and shared among devoted audiences.

As a somewhat extreme example of this trend, we might turn to one of television's longest-running titles, *Doctor Who* (BBC, 1963–89, 2005–) and note that, since a revival that coincided with the rising peak of the new 'golden age', this programme has made significant efforts to emphasise its credentials as a long-arc narrative and to create patterns of continuity stretching back across many years of broadcasts. When series 10 of the rebooted *Doctor Who* concluded with its current incarnation, Peter Capaldi, encountering the programme's first Doctor (played now by David Bradley in a studied rendition of William Hartnell's original), having battled an enemy not seen on-screen for over 50 years (the Mondasian Cybermen), we might consider that this programme had become especially invested in a process of weaving together moments from its extended history. It is also the case, in the context of this particular programme's broadcast history, that the episode makes efforts to smooth over the fracture point between the 'classic Who' (those episodes aired before the series' cancellation in 1989) and the 'new Who' (episodes occurring after the show's 2005 reboot). This is achieved by effectively placing a 2017 storyline within a pre-existing storyline from 1966, thus creating a more complex narrative that now stretches across over 50 years of screen time.

Central to this whole endeavour is the spectacle of witnessing two Doctors from opposite ends of the timeline on-screen together. The pleasure of this experience is heavily reliant upon television performance: David Bradley is obliged to deliver a meticulous execution of William Hartnell's first Doctor precisely because that version is embodied in Hartnell's particular set of gestures, facial expressions and vocal intonations. An understanding of character is intertwined with an appreciation of actor and, as a consequence, a recasting of the former must achieve a near-reincarnation of the latter. As a result, Bradley's close evocation of Hartnell's performance style is placed alongside and in contrast to Capaldi's, creating new structures of tension, resonance and depth that rely upon the notion that we are watching two Doctors and, crucially, two performances that would otherwise be separated by the passage of time. Given that, until the 2005 reboot of the programme, current Doctors encountering their former selves had occasionally been a somewhat gimmicky or even throwaway event,[1] it is significant that *Doctor Who* should make such an effort to bind together its past and present to construct a cohesive, overarching narrative that centres (and relies) upon the performance strengths of its central cast. And so, even when one of those cast members has been deceased for 42 years, as was the case with Hartnell, another actor steps up to create a facsimile performance. Taking all of this into account, we are entitled to conclude that *Doctor Who*'s efforts to construct a long-running narrative arc, and to make dramatic capital from that continuity, respond to a production climate well-established by its 2017 broadcast, whereby we could identify a significant number of television dramas that made length and depth their virtue.

In considering performance as a key facet in long-running television programmes, this chapter seeks to explore a range of texts that offer variations and distinctions within that trend. Two programmes, *EastEnders* (BBC, 1985–) and *Columbo* (NBC, 1968–78; ABC, 1989–2003), provide examples that do not fall comfortably within the parameters of television's 'new' golden age, as characterised by Lawson and Leslie, either due to their broadcast dates (although still in production, *EastEnders* began in 1985 and *Columbo* ran sporadically between 1968 and 2003) or their genre (a soap opera set in London's East End and a detective show featuring regular celebrity guest spots, respectively). The discussion then draws a relationship between these examples and two programmes that enjoy strong reputations within the 'recent crop of episodic TV production' that Martin identifies: *Breaking Bad* (AMC, 2008–13) and *Game of Thrones* (HBO, 2011–19). Offering this relative diversity of examples might help to broaden notions of what performance achievement in long-running television can be and, at the same time, offers an illustration of the ways in which performers shape their work to a programme's particular textures and contours. We might suggest that such a relationship is reciprocal as performers take an active part in forming those textures and contours, so that an evaluation of a programme's accomplishments can be strongly reliant upon the achievements of its actors. Therefore, although examples like *Breaking Bad* and *Game of Thrones* might have the potential to be bracketed together under terms like 'quality television', however contentious such definitions might be (McCabe and Akass 2007: 1–12), an attention to the work of the performers within the long span of broadcast hours can help to illuminate some of the key distinctions between those programmes' dramatic interests (in my discussion, the ways in which characters are shown to embrace or resist change, for example). In this way, the following chapter uses the breadth of narrative scale in long-running programmes to draw out notions of breadth in performance and, as a result, attempts to develop further breadth in our critical appreciation of actors and television.

Approaching 'Long' Television

Before moving on to make performance a central concern in this chapter, it is worth noting some of the ways in which matters of length and depth in television drama have been attended to more generally in various scholarly debates. Jason Jacobs and Steven Peacock, for example, describe some of the critical challenges that can be bound up in the close study of moments in television, and how these might be differentiated from the equivalent analysis of moments in film:

> [...] the expansive structure of television fictions – stretching across episodes, seasons, in series running for years and decades – complicates the place of the moment in the whole. A series alert to the possibilities of significant patterning in forms of narrative and narration may rhyme or counterpoint moments from different seasons [...] The series may allude to very early moments in its final stages, even if the mood and tone of the drama have transformed fundamentally from within [...] At the same time, the fluid consistency of a TV drama with the ability to run on forever makes mischief with the criterion of coherence. How do we judge a television work's unity if it is open-ended, changing and building across episodes, still in flux? How can we make decisive discriminations of a particular moment if its relationship to the (incomplete) whole is as yet undeclared or undecided?
>
> (2013: 6)

The concerns that Jacobs and Peacock raise here have clear implications for anyone seeking to articulate notions of achievement and quality in television. Making claims for a programme that has continued for many 'episodes, seasons, in series running for years and decades' becomes a precarious endeavour when that work may have gone through many modifications in style, tone and narrative form. And these changes will always have an impact upon perceptions of achievement. Returning to *Doctor Who*, even the most committed viewers can struggle to accommodate the budget-starved episodes of the late 1980s or the ill-conceived television movie of the mid-1990s within an overall endorsement of their show's merits. Yet, all fall under the same title: *Doctor Who*.

If changes and fluctuations can make programmes uneven and inconsistent, failing to evolve might prove equally perilous. A title that once offered a fresh reworking of television aesthetics, such as *24* (Fox, 2001–14), can gradually seem formulaic and contrived when those innovations are made to stretch across nine seasons spanning 14 years. Likewise, a title that strived to generate ingenious levels of depth in its narrative exposition, such as *Lost* (ABC, 2004–10), can see its meaning and significance evaporate steadily as events struggle to find resolution within a vast, meandering sequence of episodes and seasons. In these instances, length and depth can become potential hindrances for television programmes, threatening to undermine key achievements as they slip into repetitiveness or even redundancy.

We can identify related interests in television drama's length and depth running through the work of Glen Creeber and Jason Mittell, who formulate their arguments respectively around notions of seriality and complexity (Creeber 2004; Mittell 2015). Although each concentrate on detailing in relatively objective terms the formal composition of certain case studies belonging to either serial or complex TV (with Creeber analysing key socio-political themes in programmes and Mittell employing a methodology of poetics to consider a range of informing contexts) and so are not engaged consistently or exclusively in the articulation of achievement and quality, it is noteworthy that they nevertheless gravitate towards programmes that have frequently been regarded as examples of outstanding creative accomplishment: *Twin Peaks* (ABC, 1990–91), *The Sopranos* (HBO, 1999–2007), *Prime Suspect* (ITV, 1991–92, 1993–96, 2003–06), *This Life* (BBC, 1996–97, 2006), *Breaking Bad, Lost, Battlestar Galactica* (Sci-Fi, 2004–09), *Arrested Development* (Fox, 2003–06; Netflix, 2013–), *The Wire* (HBO, 2002–08), *Mad Men* (AMC, 2007–15) etc. We might reason that these programmes are taken to demonstrate the best examples of the length and depth Creeber and Mittell identify as a key development in contemporary television. This, in itself, might constitute a claim for particular kinds of quality and achievement. In a related strategy, both authors express a keenness to disentangle their chosen texts – and the trends they represent – from an association with the genre of television soap opera, which might otherwise offer potent opportunities in a discussion of broad narrative arcs. For Mittell, in a debate focused around melodrama more broadly, there is a need to challenge the perception that an evolutionary connection exists between the prime-time complex television texts he identifies and examples of daytime soap opera that precede them. As he explains:

> While contemporary prime time television embraces seriality in a range of ways, I contend that the specific modes of serial storytelling it employs derive less from American soap operas than from other serial modes such as comics, classic film serials, and 19th-century literature, all of which have their own connections to melodrama. But since the history of television seriality is so linked to the soap opera genre, the common assumption is that all prime time serials must be reacting to or building on soaps, an assumption that I hope to break apart here.
>
> (2015: 236)

Mittell makes a pronounced distinction in this passage and we might find reason to question in more detail whether, in fact, the relationship is *stronger* between prime-time complex television and 'comics, classic film serials, and 19th-century literature' than soap operas, given questions of medium specificity that may lead to points of commonality. Nevertheless, the concept of a range of influences impacting upon the shape and structure of something called complex TV has obvious merits, especially when taken in the context of Mittell's framing poetics approach.

Focusing on UK television, Creeber adopts a slightly different position as he builds distinctions between serial drama and soap opera. Although conceding that contemporary television serials may have a 'soap-like' quality, he contends that:

> While their connection to continuous drama is there for anyone to see, there is still a discernible difference between these serial dramas and soap opera. Unlike most soap operas these serials:
> - Are usually broadcast post-watershed (meaning that their content and characterization can be decidedly more adult and intense in tone).
> - Combine a mixture of 'flexi-narrative' techniques within a loosely defined narrative arc (meaning that elements of narrative progression towards conclusion can and still take place but within a complex exchange of narrative and character complexity).
> - Sometimes reveal a tendency towards more 'experimental' techniques (meaning that definitions of social reality can be and are increasingly called into question. This clearly includes the subversion and re-invention of genre [...]).

- Offer examples of a new relationship between politics and the self (meaning that political issues are now increasingly centred convincingly around the domain of personal and private interaction). (2004: 12)

Creeber is careful to preface his list of points with the clause 'unlike *most* soap operas' but, clearly, the distinctions he puts in place are open to contention, given that examples of soap operas might readily be found that correspond closely to the characteristics he associates with serial drama. However, like Mittell, he establishes a method for distancing his chosen case studies from soap opera and, in using a set of characteristic traits to highlight some of the progressive ambitions that might be exhibited by that corpus, Creeber begins implicitly to form claims for the particular value of serial television as distinct from soap operas.

The kind of separation that Mittell and Creeber establish, while logically argued, may serve to diminish the extent to which soap operas offer useful examples of how length and depth might be balanced creatively and effectively in a conventional narrative structure. As Christine Geraghty notes, 'Soaps constitute very difficult and exhausting texts in the sense that the duration and narrative complexity of the most successful examples make even *Lost* or *The Sopranos* [...] seem manageable' (2010: 91). Focusing simply on characters in soap opera, for example, emphasises those levels of complexity that Geraghty describes, which build and intensify as each new episode is broadcast. As a consequence, audiences can be invited to develop and maintain a particularly intricate and convoluted understanding of characters on-screen, even when those characters are engaged in apparently straightforward storylines. It is an achievement of soap opera that any episode can potentially be watched and enjoyed without a detailed contextual appreciation of past episodes that informs its current plot points. However, even casual viewers possess a background knowledge of character history built up over years. As a result, it may well be that one character's actions in a particular moment can be understood against a backdrop of previous actions – even when those actions have reached a point of resolution and do not possess an obvious bearing on current storylines. For dedicated fans of soap operas, that potential is only intensified and becomes a key source of viewing pleasure.

Overlooking the significance of length and depth in soap operas might equally blind us to the qualities of performance in programmes of this kind. Like viewers, soap opera actors have the opportunity to develop detailed understandings of their character's expanding history and, by consequence, awareness of that history. They can choose how to use this knowledge. June Brown, for example, has portrayed the character of Dot Cotton in *EastEnders* for over 30 years and has remarked that she 'always felt Dot as one of those characters who should stay the same. She's a simple creature. There are some people who the same things happen to them again and again. They never learn' (Ward 1997). While on the surface this may read as an invitation to resist viewing the character of Dot Cotton and, by implication, Brown's portrayal as especially complex (because nothing ever changes or develops for her), it is equally the case that Brown articulates specific performance choices that best represent Dot's psychological response to the increasingly complex set of circumstances that she encounters in the passage of time. Over a period of 30 years between 1985 and 2015, for example, Dot's reprobate son Nick Cotton (John Altman) returns frequently to his mother, invariably bringing fresh pain and anguish. The regularity of this charade surely presented Brown with some acute challenges in terms of representing her character's emotional and intellectual perspectives (namely, why Dot never learns from this blatant repetition). Yet, Brown's performance managed to capture Dot's psychological tension as she balanced her character's logical awareness (and, often, fear) of Nick's persistent criminality against an irrational desire to believe him somehow capable of redemption. A notable and recurring technique of Brown's was to silently search Altman's features as he spoke, often when the camera was framing her in a shot reverse-shot structure or both actors in a two-shot (Figure 3.1).

The effect was of Dot attempting to penetrate the veneer of Nick's appearance, to somehow discern the nature of this man through visual scrutiny. It was as though she was asking herself again and again what she was seeing and, in turn, we were invited to analyse her features as she gazed to ask, for ourselves, what it was that Dot saw in Nick. Consequently, through Brown's understanding and expression of her character in moments of this kind, Dot's apparent inability to learn from past experience becomes a legitimate area of interest and conjecture for viewers.

It is uncontroversial to suggest that Brown's performance in *EastEnders* is accomplished and, indeed, she has received sustained recognition for her acting achievements in the programme. A 2009 British Academy of Film and Television Arts (BAFTA) nomination acknowledged her work in an episode, broadcast on 31 January 2008, that was unique for a soap opera in

Figure 3.1 June Brown silently searches John Altman's features (*EastEnders*, BBC, 1985–)

that it featured no other actors. (Brown's BAFTA nomination was equally exceptional given that no other soap opera actor had been a recipient for 30 years.) This 'single-hander' placed all emphasis upon Brown's acting ability as she delivered a continuous monologue for the entire 28-minute episode. This shift in dramatic focus naturally effected a change in dramatic form, moving *EastEnders* temporarily away from the fluctuating tones of a conventional ensemble-cast soap opera and towards the intense confessional style of Alan Bennett's *Talking Heads* (BBC, 1982, 1988 and 1989). In the context of this chapter's interests, we might note that Brown's performance leads this development, free to expand beyond the orthodox boundaries of the soap opera format and *become* the episode, offered to the viewer in greater length and detail because Brown is able to meet and reward those intensified levels of scrutiny. It is tempting to regard the episode, and Brown's performance within it, as standing apart almost entirely from the qualities and quality of an 'ordinary' *EastEnders* instalment, a sentiment encapsulated in Decca Aitkenhead's claim that it possessed 'an understated economy that transcends the cartoonish confines of most soap opera' (Aitkenhead 2009). Self-evidently, judgements of this kind are also articulations of personal taste, whereby the episode gains value only as it moves away from the soap opera genre's 'cartoonish confines' and aspires towards the 'understated economy' of quality drama. Whether or not we agree with this limited characterisation of the genres, it is nevertheless the case that Aitkenhead's distinction, in divorcing the episode from soap opera, avoids accounting for the ways in which the single-hander could only exist due to the hours of screen time devoted to the character of Dot Cotton within the 'confines' of *EastEnders*' conventional broadcast format. More specifically, over many years, levels of depth and complexity have built up both within the writers' accumulating storylines and within Brown's depiction of Dot to the extent that a full episode could explore the character in yet more precise detail without breaking the programme's overarching dramatic style and, indeed, Brown's acting style. In this sense, and in response to an attitude that Aitkenhead's claims typify, 'understated economy' was always a feature of Brown's performances in *EastEnders*, exemplified in those repeated exchanges with Altman, so avoiding a broad, one-dimensional or, indeed, 'cartoonish' portrayal.

'One More Thing': Performance as Unpredictability in *Columbo*

Focusing on aspects of June Brown's performance in *EastEnders* brings to mind some of the ways in which soap operas can sometimes be characterised as straightforward when compared to other television genres (Geraghty 2010: 82). If we distinguish a long-running television text as repetitive, linear or simplistic, we also risk categorising performance in similar terms, thus subduing an appreciation of actors' work in managing their characterisations across years of screen time. In this way, soap operas can provide a useful foundation for thinking about performance in other areas of television that might be regarded, on the surface at least, as equally limited in terms of dramatic scope and characterisation. While not in continuous production, *Columbo* enjoyed an especially long screen life and each episode relied upon a template three-act plot structure that can be regarded as especially formulaic and limited. As the show's co-creator, William Link, explains:

> Whole first act, you see a murderer [...] committing what the character perceives as the perfect crime; second act, enter Columbo. A duel of wits ensues: the rest of the show is basically this minuet [...] between two people. At the end [...] there's the big clue that allows Columbo to catch the murderer. That's the format!
>
> (Archive of American Television, 2002)

The obvious repetitiveness of this structure might appear uninspired and unsustainable over time were it not for features like a regularly refreshed team of writers, directors, guest stars and, most importantly, the consistent acting achievements of Peter Falk in the role of Lieutenant Columbo. Falk's performance works against the potential limitations of *Columbo*'s narrative constraints, creating surprise and unpredictability in episodes where so much is already known, including the identity of the murderer. His characterisation becomes a key structuring element, devising complicated idiosyncratic patterns of gesture, speech and movement that run across the plotted directness of the show's 'three acts' structure. In this way, Falk's acting style is in tune with the central aim of his character, Columbo, to wrong-foot the complacent murderer in each case.

The episode 'Death Hits the Jackpot' aired in 1991, and by this point in the programme's history *Columbo*'s narrative patterns and Falk's performance style were very well-established. One requirement for the actor, therefore, is to avoid scenes becoming dulled by this inherent predictability. In such a scene, Columbo visits the murderer, Leon Lamarr (Rip Torn), in his jewellery store. In other hands, this could be a routine exchange between detective and suspect designed to highlight the latter's attempts to deceive the former. In *Columbo*, however, it is the lieutenant's demeanour that is deceptive as his absent-mindedness provides a cover for keen perspicacity. The scene begins with Columbo knocking on the jeweller's window and pointing to a particular item, his words inaudible through the reinforced glass. On entering, he explains that he is hurried due to an impending vet's appointment but enquires about a silver charm bracelet in the window (Columbo's search for a silver wedding anniversary gift for his wife is a recurrent plot-point in the episode, while his dog features in many seasons of the programme). Lamarr suggests a better item but Columbo insists he has to leave. However, the lieutenant has an apparent change of mind, remarking casually that he has a question about the recent murder victim (Lamarr's nephew). Lamarr shows Columbo into a glass-doored rear office and Falk begins his character's questioning before he steps into the room. This not only continues a lack of formality in the exchanges, established with the shop window introduction, but also reflects the fact that Columbo does not generally observe spatial boundaries or, indeed, social etiquette. This echoes scenes elsewhere in the episode in which he interrupts the victim's funeral, gatecrashes an impromptu wake and disrupts a jewellery auction by inadvertently bidding against Lamarr on a particular item. In this scene, Falk continues this transgressive behaviour as he begins a series of new questions while making to leave the office, standing in the doorway or walking away across the shop floor, in each case maintaining a distracted demeanour as he repeatedly looks down and taps his watch or rearranges the crumpled documents he clasps awkwardly in his cigar-bearing hand. In the conversation, Falk adopts a particularly genial tone and begins to mimic Torn's soft Southern drawl as he elongates and accentuates the vowels in words like 'mind' and 'money', as though Columbo were giving Lamarr a false sense of authority by partially mirroring his style and tone. As a result of this fluidity of movement and speech, the delivery of the trademark line 'one more thing' is naturalised and made coherent within an existing pattern of behaviour, whereas otherwise it might easily register as contrived due to its prominent repetition across episodes.

With the interview apparently concluded, Columbo leaves and Lamarr shuts the door to his office and walks over to a window, disquiet falling across his features as, on the soundtrack, an ominous string arrangement provides a clichéd reflection of the character's sinister thoughts. And, yet, he is interrupted again as Columbo appears once more at the office window, knocking on the glass just as he did at the beginning of the scene. The meaning of Columbo's knock has changed for Lamarr, of course, from an unwelcome distraction to a threatening intrusion. Falk is careful to avoid this shift becoming exaggerated as he maintains the same easy congeniality that characterised his first appearance at the shop window. And so the two meanings of the window-knocking are achieved with subtlety, incorporated into an overarching performance rhythm rather than becoming disharmonious through over-assertion. As Columbo asks his final questions, he hovers in the doorway, stares down at his watch and gesticulates distractedly with the folded sheets of paper he holds. The lieutenant is actually retrieving a vital piece of information (that Lamarr's nephew never intended to attend his uncle's Halloween party on the night of his murder because he didn't possess the obligatory costume) but, again, Falk's casual and preoccupied delivery successfully covers the impact of this enquiry, meaning that we might miss its significance just as Lamarr underestimates Columbo and subsequently gets drawn into constructing a flimsy explanation. Falk does create a degree of intensity, however, as he steps into the room and lightly allows the door to swing shut behind him by giving it a very slight tug (Figure 3.2).

Given Columbo's earlier disregard for spatial boundaries, this attempt to create an enclosed environment might register as inconsistent. Yet, Falk's action allows his character to finally 'corner' Lamarr, turning the encounter into a more focused interrogation as Columbo stands blocking the closed door. Although understated and lacking any emphatic flourish (closing the

Figure 3.2 Peter Falk presents interrogation as interruption (*Columbo*, NBC, 1968–78; ABC, 1989–2003)

door might equally be read as an act of consideration for Lamarr's privacy), Falk's performance helps to position this moment as a significant turning point for Columbo: under pressure, Lamarr lies to him. In this context, Falk's final action demonstrates that Columbo's earlier obliviousness towards spatial boundaries did not stem from a lack of understanding but, rather, forms part of an investigative strategy. At the end of this scene, he lets the door shut behind him at just the right moment, creating an enclosed setting to exert a degree of pressure on his suspect.

Paying attention to Falk's performance in *Columbo* can reveal degrees of subtlety and complexity that reward closer repeated viewing. Certainly, we can identify layers of detail and nuance that work above and beyond the basic requirements for a television detective. This is especially important for *Columbo* as so much is known already – the murderer, the victim, and the method – and so much is already familiar to us – Columbo's investigative style and the 'three acts' structure as described by Link. In a narrative structure that capitalises on the strengths of formulaic predictability, Falk is able to find renewed opportunities for character depth and diversity, even after 23 years, and consequently for creating a delicate tension between linear plot and non-linear performance.

'I Am the Danger': Performance as Metamorphosis in *Breaking Bad*

The character of Columbo is afforded sparse potential for change or development, given the closed narratives and infrequency of episodes. In other long-running programmes, opportunities to perform a character's growth and even transformation can be found more readily. Indeed, this emphasis can form the basis of a series' entire narrative trajectory. *Breaking Bad*, for example, is notorious for charting a character's transition from one identity to another, depicting the evolution of high school teacher Walter White (Bryan Cranston) into an invented drug lord alter ego, 'Heisenberg'. The tension between the two identities is played out in numerous scenes throughout *Breaking Bad* and, as a consequence, Cranston is required to balance the twin sides of his character while preserving the authenticity of each. In the sixth episode of season four, Walter's wife, Skyler (Anna Gunn), attempts to persuade her husband to give up his criminal life and go to the police for protection. The scene takes place in the couple's bedroom and, as Skyler makes her case, Cranston performs Walter's characteristic evasiveness and unwillingness to engage at home as he never turns his torso fully to face Gunn, dropping his eyeline to avoid direct contact and speaking his lines in a tense undertone that repeatedly cuts into his fellow actor's words. Walter finally silences Skyler by raising his hand, standing up from the bed and saying abruptly 'OK we're done here' as a means of closing off the debate entirely before walking across the room. With her husband turned away from her, Skyler nevertheless continues to address his back: 'Walt, please, let's both of us stop trying to justify this whole thing and admit you're in danger!' The word 'danger' provokes a changed response in Walter and he turns to face his wife, replying: 'Who are you talking to right now? Who is it you think you see?' As Cranston delivers these lines, he walks back over to the bed, to look Gunn directly in the eye and stepping forward to tower over her in a position of Walter's heavily asserted authority over Skyler. After a pause made intense by his unremitting gaze, he continues: 'Do you know how much I make a year? I mean, even if I told you, you wouldn't believe it.' Cranston stiffens his posture and measures his movements to illustrate a change of pace, a hardening of the conversation and a strengthening of his character's self-belief, but then he gains fluency as the speech builds: 'Do you know what would happen if I suddenly decided to stop going into work? A business big enough that it could be listed on the NASDAQ goes belly up. Disappears! It ceases to exist without me.' And then he doesn't even pause for a second before moving into the final, climactic section of his monologue: 'No, you *clearly* don't know who you're talking to, so let me clue you in. I am *not* in danger, Skyler. I *am* the danger. A guy opens his door and gets shot and you think that of me? No. I *am* the one who knocks.'

As Cranston delivers this speech, his voice gradually drops into the raspy lower register that Walter most regularly adopts in his alter ego of Heisenberg. He allows a sharp aggression to build in his delivery, which is complemented by the use of a series of strong, precise hand gestures to emphasise certain sections in the speech where Walter points with a flat hand at Skyler and later with a closed fist and single finger towards himself.

These transitions in performance style frame Walter's words as both impressive and intimidating, exposing to Skyler the latent violence he possesses and also emphasising his pleasure in asserting this kind of authority over another person. There is a rhythmic fluidity and incessant drive in Cranston's delivery that contrasts forcefully with the strained staccato pace of the lines delivered earlier on the bed. Walter no longer shuts Skyler out but instead patronises her, vocally bombards her, as he lectures her on the realities of his existence (his standing position allowing him to 'talk down' to her as she remains seated, creating an imbalanced power relationship that also exploits gender difference). Cranston in fact manages to contrast the two realities of Walter's existence in this scene, his home life and his 'work' life persona, bringing them into collision with one another through performance decisions and details so that, effectively, Skyler is left facing Heisenberg for a matter of seconds ('Who is it you think you see?'). Furthermore, Cranston gives this moment prominence as the scene's dramatic high point by infusing the 'Heisenberg' section of the speech with qualities that make it especially engaging for a television audience: strong and sincere, rather than subdued and evasive (Figure 3.3).

In his performance choices, Cranston succeeds in establishing the Heisenberg identity as most natural for Walter, allowing him to communicate more directly, more openly and with greater conviction. Consequently, this small moment projects a bleak future for Walter and Skyler over the following seasons, as he will become consumed by his alter ego and she ultimately will lose her husband forever. This contrast and conflict between

Figure 3.3   Bryan Cranston reveals Heisenberg (*Breaking Bad*, AMC, 2008–13)

personas hinges upon notions of the *character's* relationship to his performance: initially, Walter was required to perform the role of 'Heisenberg' the drug lord, the maintenance of which required considerable effort and control, whereas that same exertion is now required to maintain the performance of 'Walter' the family man. The depth and scale of Walter's transformation, and Cranston's depiction of that metamorphosis, from season to season complements an acute appreciation of *Breaking Bad's* equivalent depth and scale. The long arc of character development is dependent upon the length of screen time devoted to portraying that journey. It is no surprise, then, that numerous fan-made online videos make Walter White's evolution their central interest, condensing the span of his descent into minutes in an effort to illustrate patterns that otherwise emerge only across many hours of television. These videos attempt to compress the elongated transition that Cranston achieves in his character, necessarily bypassing the subtlety and nuance of his long-spanning portrayal in an effort to cluster together the pinnacles of that performance.

Feeling Good: Performance as Obdurateness in *Game of Thrones*

Picking out these features in *Breaking Bad* may risk inviting the assumption that performance achievement in broad-arcing television narratives must always be related to the successful realisation of character development, growth or even transformation. However, it is also the case that actors can find levels of depth and complexity in characters that are defined by their resistance to change. We might think again of June Brown's portrayal of Dot Cotton here, and the ways in which her performance encompasses the notion that the character's failure to properly learn from and move beyond past experiences stems either from an inability or an unwillingness to change, or a combination of both. We can look to other styles and genres of television drama for examples that explore and develop this tension. Cersei Lannister (Lena Headey) has consistently been a central player in the convoluted plotlines of *Game of Thrones* and, as a consequence, she frequently responds to a world that shifts and reshapes around her. We might contrast this with *Columbo's* fictional world, in which the lieutenant constitutes the unpredictable element in an otherwise fixed and even formulaic diegesis. In *Game of Thrones*, developments can be extraordinary, illustrated by the extent to which murder has been a constant feature in Cersei's life: she arranged for her husband to be killed, her father was killed by her brother, her two children were killed by enemy hands, not to mention the many deaths that she has either orchestrated or witnessed as she strives to gain power in her Kingdom of Westeros and, later, across the Seven Kingdoms. In season five, Cersei is imprisoned and brutally tortured by a malicious religious sect, the Sparrows, who have taken control of the city in part due to her own political machinations. This torment culminates in her being made to walk naked through the streets to atone for her sins. These scenes of Cersei's torment and humiliation represent the lowest emotional and psychological point for the character, offering the clearest potential for personal reflection and growth. To a degree, this does happen: in the tenth episode of season six, Cersei responds by taking revenge on the Sparrows and other assorted enemies, blowing them all up in a city temple.

From a balcony in the Red Keep (her castle residence), Cersei watches the smoke and flame rise from the explosion. She breathes deeply as a smile of satisfaction forms on her lips, raises her glass of wine and holds it suspended momentarily, as though in mock toast to the ruined temple, before taking a slow sip, turning and leaving. We next see Cersei in a dungeon room or, rather, we see her hand as she pours a glassful of red wine over the face of Septa Unella (Hannah Waddingham), a member of the pious female clergy serving the Sparrow, and Cersei's chief torturer during her captivity. A cut to Unella's hands reveals that she is manacled. A further cut to a wider shot introduces Cersei's voice on the soundtrack as she orders 'Confess. Confess. Confess', before walking into shot and tipping a whole carafe of wine over Unella's face. A reverse-angle shot reveals Cersei, impassive, functional, as she performs her torment of a former captor. The reciting of 'Confess' mimics a line Unella repeated when she held Cersei prisoner and its use here is clearly intended to emphasise the reversal of power that has taken place, just as the use of wine as an instrument of torture replaces the water Unella once withheld from her prisoner: an aptly corruptive, intoxicating liquid that the septa disapprove of strongly. Cersei continues: 'Confess: it felt good, beating me. Starving me. Frightening me. Humiliating me. You didn't do it because you cared about my atonement. You did it because it felt good. I understand. I do things because they feel good.' Headey softens the delivery of these lines, contrasting them with the strong hard edge of her preceding orders to 'confess'. She places a fragile emphasis on the words 'beating', 'starving' and 'frightening', leaving pauses between them before rushing 'humiliating'. The measure of this delivery conveys a sense of naturalistic speech, attaching thought and meaning to the words, as though Cersei were reliving the vivid experience of each cruelty as she describes it, conveying her past vulnerability and losing her control very slightly as she pushes out the most painful recollection: 'humiliating'. Headey's vocal performance complements her character's attempts to explain Unella's actions as base human satisfaction rather than religious duty. A heavier, breathier emphasis is placed on 'I understand', as though Cersei were exhaling in relief that these two women share sensations and motivations, somehow intimate through their mutual love of cruelty and honest with each other now. Headey never allows her focus to leave the face of her fellow actor as she delivers these lines, maintaining Cersei's attempted emotive connection with this woman as she relives her ordeal, describing their sadistic bond and, at the same time, conveying the strong impression that she has Unella 'in her sights': that her intentions are nakedly malevolent.

Cersei continues her speech, circling the bench to which Unella is shackled:

> I drink because it feels good. I killed my husband because it felt good to be rid of him. I fuck my brother because it feels good to feel him inside me. I lie about fucking my brother because it feels good to keep our son safe from hateful hypocrites.

This is actually the confession that the Sparrows craved, useless to them as it is delivered now. And embellished with sacrilegious action: as Cersei moves around Unella, she first gently strokes the arm of her victim before tugging at the ropes that bind her, and then letting her hand rest on the captive woman's stomach and sliding it down her inner thigh, causing Unella clear discomfort as she spasms in futile resistance. Headey's performance of these gestures in one sense emphasises Cersei's power over Unella, that she can place her hands wherever she chooses, but it also draws the prisoner into the licentious behaviour being described to her, forcing her to feel echoes of the sensations being described graphically to her. Occurring within Cersei's groping of her victim, the tugging of the restraints therefore introduces a flicker of playful sadomasochism to the encounter, briefly altering the meaning of Unella's bondage from religious martyrdom to a game of sexual gratification, which, like a confession of incest, would outrage the septa. As Cersei describes protecting her son from 'hateful hypocrites', Headey stands upright, fixes her resolute gaze again on the captive and allows her features to harden, the smile now faded from her lips, revealing momentarily the steadfastness of Cersei's protection and also making ominously clear that she identifies Unella as one such hypocrite (Figure 3.4).

Moving around the bench, Cersei takes hold of Unella's resisting hand and leans forward to confide in a hushed tone: 'I killed your High Sparrow, and all his little sparrows. All his Septons, all his Septas. All his filthy soldiers. Because it felt good to watch them burn. It felt good to imagine their shock and their pain.' Framed in intense close-up, Headey grits her teeth and curls her lips into a faint snarl as Cersei describes her act of mass murder. It is a performance technique the actress employs throughout the seasons of *Game of Thrones* to illustrate her character's disgust or rage. Here, the expression reveals her abhorrence for the men and women she describes and also the passionate fervour she experiences in imagining their terror, the 'good' feeling she refers to, gritted teeth containing and conveying her deep visceral satisfaction. This strength of emotion is complemented in Headey's vocal

Figure 3.4 Lena Headey conveys Cersei's ominous resolve (*Game of Thrones*, HBO, 2011–19)

delivery as she allows her tone to waver and tremble, cracking with the intensity of Cersei's sadistic gratification. We cut to a wider shot as her speech continues: 'No thought has ever given me greater joy. Even confessing feels good, under the right circumstances.' Headey lightens her expression again for the delivery of these two lines, opening her features out into a countenance of ironic virtue and innocence for the first sentence, exhaling deeply as though moved spiritually in a mocking reference to Unella's cruel piety. And then she lets her gaze fall slightly, giving a short laugh before offering the second sentence as candid afterthought, contrasting with the careful crafting of her earlier speech. This change of style in Headey's performance drains some of the intensity from the scene, a notion complemented as Cersei softly places a hand on Unella's face. But this merely provides a brief respite from one ordeal before a new one takes its place: Cersei prepares Unella for a long, drawn-out future of pain and introduces her to Ser Gregor Clegane (Hafþór Júlíus Björnsson), a giant undead guardsman, who will administer that suffering.

Headey's performance of Cersei in this sequence involves a range of different textures and tones, building a sense of her character's psychological depth and complexity. In the course of the exchange, Cersei moves between ruthlessness, sympathy, vulnerability, strength, sarcasm, honesty, innocence, guilt, sadism, pleasure, perversion, relief and tension. The character's formidable ability to fluently balance these different states is also the actor's achievement as Headey manages this diverse and affecting performance. And yet, although the act is multifaceted, at its core is Cersei's resolute refusal to adjust or adapt, whatever circumstances may arise. She commits herself defiantly to the twin hedonistic pursuits of pleasure and power, entwined as they are for this woman, that have propelled her forward through each densely structured season of the programme. Rather than causing her to adjust her trajectory, Cersei's imprisonment and torture at the hands of the Sparrows has only strengthened her defiance, fortifying her convictions and, ultimately, making her more secure in her established behaviour. It cannot even be claimed that the trauma creates a monster. Instead, Cersei returns to her life with reinforced confidence. In this way, the character is made constant and unyielding in the programme's complicated, interwoven and evolving narrative structures. She will not develop and she will not change. Indeed, this scene plays out as her manifesto for staying the same. Headey's final achievement, then, is ensuring that this focused consistency does not become limiting: that Cersei does not become the standard caricature of evil committed only to being evil. The actor's dexterity in combining different tones and styles of performance in this one sequence creates an engaging and challenging portrayal that rewards the careful writing and composition of the scene. Thus, sameness is made a guiding and enticing facet of Cersei's character.

Conclusion

In many ways, Cersei can be regarded as a fixed point in the ever-shifting events of *Game of Thrones*, given that she commits so emphatically to an unwavering set of motivations and behaviours. We might equate this notion to an evaluation of performance in 'long' television more generally. As series run over years and years, and as we engage with multiple texts simultaneously, the challenge of navigating our experiences and understandings becomes acute. Simply writing about a few short examples of performance in this chapter has, for example, involved watching many hours of television content in part to check that claims are at least made in an appropriate context. As a consequence, there is a task in finding useful points of engagement: elements that are discernible and definable in texts that are so expansive and layered. One strategy is to focus on those small moments in which we take performances to feature significantly, both as illustrations of a programme's thematic interests and as examples of an individual actor's achievement. It becomes apparent, then, that drawing critical attention to performance in these moments is a way of holding onto and holding up a piece of something that is otherwise intrinsically vast and mutable. To return to points made towards the beginning of this chapter, it is difficult to talk meaningfully about the achievements of a programme by referring only to its title, because that title may encompass any number of fluctuations in style, tone and quality. Isolating performance achievement in specific moments is an available method of sharing more precise evidence for the claims we wish to make, and for attending to the structural intricacies of texts that grow in scale and depth with every new season. It is crucial, however, that the particularities of performance in television are preserved in such debates. The respective achievements of Brown, Falk, Cranston and Headey, for example, are not uniform and possess different shapes and textures according to the programmes in which they occur. Television does not provide us with fixed criteria that we can use to evaluate performance achievement *in general.* Even this chapter's arguments, while focused fairly narrowly on drama, have nevertheless been required to trace out certain key distinctions and divergences that exist between performances. If the discussion were broadened to incorporate a wider range of genres and formats, no doubt the boundaries would again expand accordingly. In this way, performances in television place a certain demand upon the critical viewer, requiring us to match any evaluative criteria to the evidence provided on-screen. As we begin this task, we are asked to replicate the kind of careful dexterity and responsiveness that television performers achieve, so aligning criticism with practice.

Notes

1   In 1993, for example, the 30th anniversary of the programme was marked by an episode, 'Dimensions in Time', included as part of the BBC's Children in Need charity broadcast that year. Shown in segments, the episode brought together all five surviving Doctors for the first time in order to stage an ill-conceived 'crossover' with the BBC's flagship soap opera *EastEnders* (1985). Viewing the realisation of this concept on-screen, it seems fair to conclude that no one involved in the production was burdened by an especially strong commitment to *Doctor Who*'s dramatic integrity.

4

Faces of Allegiance in *Homeland*: Performance and the Provisional in Serial Television Drama

Elliott Logan

One noteworthy aspect of *Homeland*'s (Showtime, 2011–) first season is how deeply it internalises a set of interpretive issues raised by television as a medium for performance. This is the case for season one in particular because of the extent to which its early episodes centre on the watching of closed-circuit television surveillance. The subject of that surveillance is US Marine Sergeant Nicholas Brody (Damian Lewis), who was presumed killed in the 2003 invasion of Iraq, but now, eight years later, has been discovered alive in al-Qaeda captivity and is returned home to be feted as a war hero. In the series' prologue, however, set ten months before Brody's rescue, CIA agent Carrie Mathison (Claire Danes) receives intelligence from a source in Baghdad, who tells her (or so she claims) that 'an American prisoner of war has been turned'.[1] Upon Brody's repatriation, Carrie suspects that he is the turned prisoner planning an attack against America. Unable to convince her CIA superior Saul Berenson (Mandy Patinkin) that Brody warrants surveillance, Carrie secretly installs cameras and microphones throughout the Brody house, and pursues her own illicit programme of watching. In the clearest emblem of *Homeland*'s self-consciousness, Carrie's monitoring of Brody is repeatedly presented as a reflection of television viewing. As Stephen Shapiro observes, for example, there is a small '*mise en abyme*' in episode two ('Grace', season one, episode two), where Carrie lies on her couch at night watching Brody as he lies on his own couch watching basketball, a mirroring within the fiction that in turn reflects our own spectatorship of it (2015: 155).

Such self-reflexive moments often mirror not only the basic fact of our television viewing as such, but also the specific kind of response to televised performance that *Homeland* itself demands. Writing about his experience of the pilot episode, Jason Jacobs recalls being

> struck by how similar Carrie's intense scrutiny of Brody's words, gestures, and movements was to my own practice of studying television: sitting up close to a big screen, notebook nearby, pen in hand, rapt, homing in on the slightest details – dare I say *clues?* – in order to assess a performance.
>
> (2011; original emphasis)

The central problem facing Carrie – and, through our involvement with her, us as well – is that of reading Brody's actions and gestures, of somehow accurately linking them with his intentions, and in doing so reliably discerning who he 'really is': simply a traumatised war veteran alienated from domestic life, or a treasonous sleeper agent bringing the War on Terror back home and turning it against the nation he has sworn to defend? The question of what Brody is 'up to' – how we are to understand what he is doing and what is moving him to do it – is one that the series fairly quickly turns back on Carrie as well, largely in light of the bipolar disorder she keeps hidden from the CIA and which brings into doubt the soundness of her judgements and actions. Also like Brody's, Carrie's behaviour is clouded by a shadow of deceptiveness; trained as a spy, her craft is to dissemble while manipulating trust. A reliable reading of Carrie's actions is then made even more difficult because her capacity for calculated performance is less than total. Far from a figure of machine-like inscrutability, Carrie appears driven by obscure inner feelings over which she seems to exert weak control. These characteristics are made vivid in Claire Danes' face, the actress adept at rapid, extreme shifts of expression, able to contort her mouth and eyes as though in the grip of a consuming force, heightening the uncertainty of what intention or purpose can be read into her character's actions and gestures. This shifting indeterminacy of intention is also a live issue for Claire Danes and Damian Lewis as they perform Carrie and Brody. As Jacobs has observed, actors in long-form television drama will often not know where their character is 'coming in to land', what the character's plans or intentions at any one moment will end up amounting to (2012: 53). In key parts of *Homeland*, this means that Danes and Lewis must to a degree act without certitude. Their success in the roles depends on their gestures and expressions being able to hold in suspension a delicate balance of potential meanings – neither playing a certain card too firmly, nor settling into blank ambiguity or contradiction – while at the same time allowing for the significance of their performances to be credibly recast in the light of the story's future unfolding.

The artistic and interpretive challenges posed by the performances in *Homeland* are therefore considerable. My aim in this chapter is to show how the series internalises those challenges in ways that have bearing on the aesthetic interest of serial form in television drama. It is centrally through its self-conscious treatment of performance, I will argue, that *Homeland* explores the opportunities for significance that its seriality makes available. Specifically, the series uses the conditions and attendant difficulties of reading filmed performances to thematise the provisionality of meaning in serial drama – that with the serial unfolding of the fiction over time, artistic intentions and spectatorial judgements become susceptible to revision, to being retrospectively altered, or discovered as somehow false or mistaken.[2] In what follows, I specifically want to illuminate a relationship between provisionality in serial drama and *Homeland*'s fascination with the innerness of Damian Lewis. It is through the inner mystery of his face, I will argue, that the conditions of provisional and retrospective meaning in serial drama are made to speak of the peculiar, corporeal grounds on which stand our deepest commitments and allegiances, rendered here as having a grip at once deeply ingrained and strangely fleeting, fragile. In dealing with these aspects of *Homeland*, I will be treating performance as both a stylistic element and a crucial thematic subject of the series, explored not only through the work of the actors but also in a wider range of formal designs. My points will sometimes be made through description of the fictional happenings without explicit reference to the agency of the actors; in the case of *Homeland*, the fictional events themselves afford substantial consideration of performance in television. I am interested in the series' handling of performance less as a lens onto acting in television as such, and more as an opening onto larger questions concerning the aesthetic stakes of serial composition in television drama.

Accumulation and Provisionality in Serial Form

Crucial to those stakes in *Homeland* are the temporal dimensions of seriality. That the series' reflections on television as a medium for performance relate to issues of time and serial form is made clear in two brief scenes from two early episodes. The first season's second episode ('Grace') opens with a flashback to Brody's captivity. Earlier flashbacks in the pilot episode revealed Brody as a liar, contradicting the account he had given of his imprisonment, specifically that he had never met the terrorist mastermind Abu Nazir (Navid Negahban) and that he had overheard but did not witness the bashing death of his fellow prisoner, Marine Corporal Tom Walker (Chris Chalk).[3] Not only do we see Brody receiving Nazir's comfort, but in the pilot's closing sequence we also see him mercilessly beat Walker's face with his own fists, and then, distraught, accept consolation from Nazir, his sworn enemy. Episode two's flashback shows Brody under guard in the desert as he digs a grave, into which Walker's body is dumped without ceremony. While Brody sings the 'Marines' Hymn' with increasing strength and defiance, the muzzle of a handgun is pressed against the back of his head, Brody wincing in anticipation of the gunshot. As it rings out we cut to Carrie waking in fright on her couch, where she has fallen asleep watching Brody on the monitor. For a fleeting moment, the flashback is reframed as Carrie's dream. But her shocked gasp overlaps on the soundtrack with Brody's scream coming through her speakers, which we now recognise as the source of Carrie's fright; the image cuts to Brody in his own bed, gripped by terrified confusion as he emerges from his nightmare. The scene's reversals of perspective and overlaps of subjectivity evoke Carrie's growing but obscure sense of one-way intimacy with Brody, of which she herself may not even be aware.

This idea is further developed in the first sequence of episode four, 'Semper I' (season one, episode four). We open on images of daily routine, which for Carrie are shown to be structured around, and mediated through, the surveillance monitor. Dressed for the office, Carrie places her coffee and breakfast on the table as she resumes her work of watching. Her own habituation to routine is then reflected in the subject of her surveillance, as we see Brody's wife Jessica preparing breakfast in the kitchen, and through the diegetic speakers we hear what turns out to be a running shower. Carrie has arrived back at her post in time to see Brody step from the shower as he dries himself with a towel and then proceeds to dress for another day giving televised speeches to promote the war. When Brody stands before his wardrobe in briefs and a white undershirt, Carrie narrates in advance his adherence to Marine Corps uniform protocol. 'Service A's today, Marine – shirt first,' she says, and then, 'Green pants.' Each is called a beat before Brody takes the items from their hangers. When he searches for his missing necktie, Carrie anticipates and answers Brody's question to Jessica before it is even asked:

'It's on the back of the bathroom door,' she says. To Carrie's amusement, Jessica echoes her words only moments later. In terms of its bare, scripted action, the scene might be taken to indicate Carrie's deepened attunement to Brody, insofar as she is able to predict his actions and therefore, to a degree, what he intends to do.

The opening sequences of 'Grace' and 'Semper I' resonate with conceptions of serial drama's interest as a medium for characterisation and performance, specifically in regard to the form's temporal aspects. In their rhyming echo, the scenes measure Carrie's increasing alignment with Brody over time. (On the same morning that she watches Brody dress, we learn that one month of surveillance has passed; on reporting to Langley, Carrie is reminded that her four-week warrant will expire the next day.) The opening of 'Semper I' draws on our memory of the earlier opening from 'Grace', both episodes beginning with the start of a new day as viewed by Carrie through her surveillance monitors. But where the 'Grace' scene is marked by the disorienting violence of Brody's nightmare, the example from 'Semper I' is characterised by calm routine, evoking the accrual of regular repetition as a basis for firm familiarity. In these terms, the latter presents an image that accords with Philip Drake's description of the part played by performers in the interest of serial drama. Seriality, writes Drake, makes available the '*accumulation*' of an actor's expressive performance work, and 'the familiarity one builds in the repeated viewing of that performer over a significant duration' (2016: 8; original emphasis). In this way, serial drama affords 'the ability to illuminate the accumulation of character knowledge through performance' (2016: 13).

However, details of performance in the two *Homeland* openings complicate any sense of straightforward accumulation. Consider how, in 'Grace', our briefly held sense that Carrie is becoming psychologically 'linked' with Brody is achieved by a perceptual trick. The scene exploits settled expectations of dream sequence convention to overturn assumptions of privileged access to character interiority. As we cut to Brody in his bed, the camera swiftly pushes in towards Lewis as he sits up violently, his eyes wide with panic, darting from one side to another in search of escape while his rigid arms and shoulders appear paralysed, holding him in place. Woken by Brody's scream, Carrie looks to the monitor to discover its source. But in Lewis's performance of Brody's terror there is no such thing to be found – only the image of a man lost in confusion and fright, overwhelmed by an obscure sense of threat that has no intelligible outer object but which resides everywhere within. And in the example from 'Semper I', the certitude of Carrie's pre-emption is coloured by hints of unintelligibility. The knowledge Carrie demonstrates by predicting Brody's actions is that of habit, the observation of a codified military procedure. Akin to the superficiality of trivia learned by rote, Carrie's 'knowledge' hardly counts as understanding another – 'knowing their mind' in the way she needs to know Brody's. When Brody thinks he sees Abu Nazir in the bathroom mirror, it appears to Carrie as only another instance of his inexplicable terror at a looming threat which remains, to her, invisible. Carrie is right when she notes down the words 'Another hallucination?', but despite her time watching Brody she cannot make any meaningful sense of what she sees. The opening of 'Semper I', together with that of 'Grace', thus brings into question the view of serial drama's duration as simply allowing determinate knowledge of character interiority to accumulate.

Instead, across season one, the secretive innerness of Damian Lewis's face is made a fixture of *Homeland*'s interest in seriality as a medium for the provisional and retrospective. The drama's patterning repeatedly invites us to look back upon the significance of moments that rested on a particular understanding of his facial expressions and to revise our sense of them. On his arrival at Andrews Air Force Base, for example, as Brody crosses the tarmac to the podium at which he will deliver a speech, we are shown him making his way through a gauntlet of reporters and photographers while a Marine band performs a loud brass number; cameras flash and the pomp of the music blares, Lewis playing a number of sharp flinches in response. This is a familiar convention, the traumatised veteran disturbed by a welcome ceremony which in its moments of violent bombast echoes the shocks of war. But a darker lining to Lewis's expression is subsequently suggested by the flashback that opens the following episode. As discussed, 'Grace' begins with Brody digging Walker's grave as he sings under his breath the 'Marines' Hymn'. In his faltering voice, we might recognise the tune of the brass music that accompanied Brody's homecoming at Andrews Air Force Base, his 'welcome back' thus returning him to the horror of his captivity. A different reading is then made available, his discomfort a sign not of simple reluctance to face such public attention, but as the expression of an interior encounter with a shameful, secret part of himself he might wish to avoid or altogether deny.

Homecoming, Mirrors and Reflection

Homeland's interest in Lewis's face, along with a key dimension of its significance, is declared earlier in the pilot, before Brody returns to the United States. In the briefing scene at CIA headquarters during which Brody's discovery is announced, a video is shown of his rescue. Here we get our first view of Damian Lewis as Brody. The character having been stunned by a flash grenade, he is hauled onto his knees by the American soldiers, his face obscured by dirt and by his overgrown hair and beard. Brody mutters a phrase in Arabic, and then repeats the same in English: 'I'm an American.' A ripple runs through the briefing room audience; in the guise of an enemy terrorist fighter, this man reveals himself to be, 'on the inside', American. The briefing scene ends on the video of Brody's rescue frozen in a close-up of a bearded and bedraggled Damian Lewis. We then dissolve to another close view of Lewis, one not mediated by a screen within the fiction but by the steamed-up bathroom mirror in which his face is reflected. The treatment of this mirror scene is rich in thematic resonance. Now safe and secure at Ramstein Air Base in Germany, Brody has been able to shower, to begin shedding the detritus of his captivity that had clung to him as though a second skin; we appear to be seeing the first steps in what will be a long process of restoration, of trying to become again the person he was. Brody stares at his bearded reflection in the mirror, and Lewis presents a picture of transfixed stillness as he gazes into his own eyes. Where the video of Brody's rescue raised the indeterminate, potentially deceptive relationship between inner and outer, here that matter is further deepened and complicated – in addition to issues of character interiority, the mirroring also brings out the always lurking question of the actor's own subjectivity. Within the fiction, it seems, Brody sees himself. But this is not in any straightforward way what we are shown. We are presented with an image in which the English actor Damian Lewis stands before a mirror and is filmed as he acts the part of the American Marine Sergeant Nicholas Brody (at this point of the story already shadowed, for us and for the actor, by the suspicion he is not who he appears to be).[4] Who, then, does Lewis see, as he watches himself pretend to be someone else? Arguably, he sees what we all might as we catch our reflection in a mirror or a passing window: the face and body as a theatrical object that is inseparable, but somehow strangely 'apart', from our own inner sense of self.

In the scene's next moments, thoughts of theatrical self-objectification are related to issues of agency and the purposiveness or meaning of bodily action. While Lewis conveys a state of fixation as he gazes into his own eyes (performing Brody's gaze at himself), the stare appears strangely empty. It is not one of intent, but is rather deadened, or somnolent. There is an echo in Lewis's eyes of what Robert Pippin notes in the archetypal performances of film noir. 'The actors are portrayed doing things,' he writes, 'but as if hypnotized, dazed, or sleepwalking, as if they are all trying to portray what something like a kind of "passive agency", what agency but not on the reflective model, *looks like*' (2012: 17; original emphasis). The sense that in this reflection we see the look of a hypnotic – of a person in a puppet-like state – is then heightened by the cut to a close-up of Lewis's hand, as it takes up a pair of hair scissors and begins to trim his beard in slow, heavy snips. The fragmentation of the actor's body by camera and editing provides the appearance of a disembodied hand acting on its own autonomy, feeding our sense of a person separated from the face he sees in the mirror. This suggestion of hypnotism, disembodiment, and passive agency is further deepened if we consider *Homeland*'s source not only in the Israeli television series *Hatufim*, but also in John Frankenheimer's 1962 film *The Manchurian Candidate*. (That film is invoked by *Homeland* when, from his first televised speech, Brody is picked as having the charisma of a potential political aspirant, and in episode ten ('Representative Brody', season one, episode ten) is approached with an offer to run for a safe seat in the House of Representatives.)

The hypnotic, disembodied qualities of Brody's agency are accentuated again in another mirror scene which, following shortly after the first, depicts the completion of his cosmetic transformation while recuperating in Germany. Now sitting in a barber's chair, Damian Lewis's face is shaved clean, his hair cut in the neat style of a Marine Sergeant (Figure 4.1).

Not only is Brody returning alive, he is also returning *as he left*, as though outwardly unchanged. Standing behind Brody is the barber, who is visible only in the form of a pair of hands, which in an abrupt insert shot are shown as they pick up a can of shaving cream and then – at the very edge of the wide shot of Brody before the mirror – delicately nudge the foam against the hairline on Brody's nape. Damian Lewis's staring eyes here seem more enlivened than in the earlier mirror scene, now providing an intense evocation of a seeing mind within, but one divorced from the image of purposeful bodily activity supplied in the barber's precision handiwork. Highlighting the issue

Figure 4.1   The seeing mind divorced from an image of bodily intent (*Homeland*, Showtime, 2011–)

of theatricality and self-division that is inherent to Damian Lewis's work as an actor, the two mirror scenes thus make especially vivid the sense of a split between Brody's mind – his inner subjectivity and intentions – and the outer corporeal surfaces of eye, face, and gesture.

These early mirror scenes are given a further turn later in the first season, one that provides a critical light by which to read a central ambiguity of the season's final instalment, 'Marine One' (season one, episode 12). In the penultimate episode, 'The Vest' (season one, episode 11), Brody's allegiance to Nazir is cemented in horrific fashion when, under the cover of a family holiday in advance of his Congressional run, he visits a tailor in Gettysburg who has made for him a suicide vest. Brody's mission then unfolds in the season finale. As Vice President Walden (Jamey Sheridan) declares his run for president with Brody at his side, a sniper attack by Walker (whose death, it turns out, was faked) forces Walden and a host of other high-ranking political and military figures into an underground bunker, any security screening impossible in such an emergency. Brody is hustled along with them, wearing the explosive suicide vest beneath his Marine Corps service uniform. Summoning all his sense of purpose, Brody approaches Walden and flicks the switch – but nothing happens, the vest's crucial wiring having come loose in the chaos of Walker's attack. An increasingly fraught Brody then repairs the device in a bathroom stall and returns to his target. But in the instant before he can once again pull the trigger and detonate his explosives, he is interrupted by a phone call from his daughter Dana (Morgan Saylor), alerted to Brody's plan by Carrie, who has recently fallen into the grip of a manic episode. In the words of Jason Mittell, Dana's call 'inspires him to abandon his plan, as he realizes what his suicide attack would do to his wife and children' (2017: 173). Replacing the red plastic cap that guards the detonator from accidental activation, a shaken Brody promises Dana he will return home.

We might explain the writers' choice to save Brody from his mission by citing the conditions of drama production in an industrial context where ongoing series are the norm. Damian Lewis was a star of the show, and so his character's death may have posed a potentially fatal disruption to the format and its interest to viewers, jeopardising hopes of a long, commercially successful run. Even so, what needs to be accounted for is *this* way of keeping the character alive and the actor on-screen for future episodes. Mittell's above-cited summary represents an interpretation of this choice that is unsatisfying when considered in light of the performance details highlighted in this chapter.

To neatly explain Brody's refusal to destroy himself as clearly motivated by a newfound sense of familial love, we must overlook much of what we are shown of Brody through Lewis's performance, especially in the mirror scenes that introduce the actor's inhabitation of his character. I have already pointed to the sense of disembodiment, hypnotism, and self-division that characterises Lewis's presence before the mirror. These qualities express a sense of Brody's weak or passive agency, and so bring into question the relationship between the character's actions and his intentions. But a subsequent episode, which ostensibly explains Brody's terrorist allegiances, further brings into question what could have been meant *by the actor*, Damian Lewis, as he performed Brody's encounter with his own reflection in the early scenes of the pilot episode. In 'Crossfire' (season one, episode nine), a series of flashbacks show Brody, five years after he was first captured, being received at Nazir's compound in northern Iraq, where he is to teach English to Nazir's youngest son, the ten-year-old Issa (Rohan Chand). Brody and Issa develop a firm bond over time, as though through this young boy Brody might retain some sense of contact with the memory of his own children. But Issa – along with

82 other classmates – is killed in a drone strike ordered by Vice President Walden, who was at the time the Director of the CIA.

The significance of these flashbacks, however, extends beyond their expository function. When Brody arrives at Nazir's compound, he resembles the figure from the rescue video in the pilot episode, a man reduced to an almost animal state of sheer survival. He is allowed to bathe, his beard is trimmed with scissors and his face shaved clean, his hair neatly cut (Figure 4.2).

Before his bath, Brody looks at himself in the mirror for what might be the first time since his capture. It is a moment that raises the question as to how he sees the person staring back – whether he recognises himself in those eyes, or finds in his own face that of a stranger. Cleansed in Nazir's bath, restored to an image of the man who so many years ago left for war, Brody is here not returned to the person he was but rather remade as the one Nazir desires him to be. Our thoughts might be cast back to the shot of Damian Lewis in the pilot episode, staring into his own eyes in the mirror, performing Brody's reflection on himself as he returns home. That stare is then cast in a changed light; seeing himself again become the Nicholas Brody who left for war, the character also watches himself undergo once more his transformation into an agent of the enemy he went to fight. However, this

Figure 4.2 Brody's transformation at the hands of Nazir (*Homeland*, Showtime, 2011–)

retrospectively available aspect of the scene's significance was likely not meant by Damian Lewis in the delivery of his performance – the actors on the first season of *Homeland* were given the script of each episode only days before its filming would begin.[5] A sense of the provisional and uncertain is thus made part of the shot's very texture, as an image of an actor whose expressions and gestures come to mean something he did not intend – could not have intended – when he performed them. Placed so early in the pilot episode, the two mirror scenes thus teach us how to read Lewis's performance and so to understand *Homeland*'s depiction of Brody: as someone who inhabits an impossible tension between competing purposes, and who does not necessarily know why he is moved to act as he does.

Ultimate Convictions

In 'Marine One', while his thumb hovers over the detonator of his suicide vest, Brody listens to his daughter as she pleads with him to promise he will return home safely; it is a promise he makes and keeps. What is at issue is the significance of this choice, not only made by the character, but as performed by the actor and woven into the wider fabric of the series' formal design. Treated simply as a plot event, this resolution might be read sentimentally, Brody having been rescued by the power of familial love over the promise of meaning achieved in a public spectacle of vengeful death. But this would be to forget that not long before he speaks with Dana, we see Brody flick the detonator's switch, the decisiveness and finality, the absoluteness of his choice expressed in the sound mix by the sharp, resolute 'click' that resonates at the moment of intended death, which by sheer luck (and writerly intervention) does not come. Brody shows that he has what it takes, that he is sufficiently committed to his terrorist cause that to see it through he would destroy himself, betray his oath to his nation, and forever ruin and stain the remainder of his children's lives and their memories of him. In a number of scenes across both 'The Vest' and the earlier parts of 'Marine One', we are shown Brody implicitly farewelling his wife and children, attempting to leave them with memories that would let them make sense of his actions in the way he wants them to be understood. (This is the other secret purpose of the family's trip to the Civil War battlefield at Gettysburg, and these moments are saturated in a sense of meaning that for Brody's children is only provisional, will only really make sense later, but then surely not in the way he intends or hopes.) As his mission proceeds, Brody appears increasingly agitated as he approaches the decisive moment. In contrast to Chris Chalk's unfaltering lethality of purpose as Tom Walker, Lewis interprets Brody's sense of mission as conflicted. But when Brody dresses in the suicide vest, he is well aware of that mission's terrible cost, and he subsequently proves it is one he is not only willing to accept for himself, but to also impose on those he loves. The ambiguity of Brody's two central choices in 'Marine One' lies in their apparent contradiction of one another. The sentimental reading of Brody's survival thus has the problem of integrating what we are shown of his ultimate commitment to self-destruction in the name of a terrorist cause, one that is held with a clear understanding of how his actions will devastate the family he leaves behind.

The contradiction is meaningfully reconciled by the treatment of Damian Lewis's performance in the moments before both the initial detonation attempt and Brody's subsequent choice to continue living and to return to his wife and children. As Brody advances towards Walden, the detonator in hand, Lewis is framed in a tight close-up, to the almost total exclusion of the space and the people around him. He stares straight ahead as Brody steadily closes on his target step-by-step, face and eyes unmoving, an echo of his somnambulant stare from early in the pilot episode. Our sense of Brody's trance-like remove from the world is heightened by the sound design that reduces the background noise of the room to a distant hum, while Damian Lewis's breathing is loud and close, which might register as Brody's heightened sensitivity to the inner thrum of his own body. In contrast to any suggestion of attunement, however, the sound of breathing is set apart from the face and body on the screen, out of scale and time with the actor's visible intake of air, the dubbing of Lewis's breath not smoothly integrated with the image but instead set at odds with it. Our view of Brody preparing to make his ultimate choice is then intercut with a flashback to Brody's training by Nazir, the actor Navid Negahban equally framed in a tight close-up as he looks into the lens, as though Nazir is staring directly into Brody's eyes. 'There is nothing left to think about,' he intones. 'Clench your teeth. Say the Holy words. Remember Issa.' Our access to Brody's subjectivity as he prepares to detonate his bomb is thus marked by a sense of the character's self-remove, dissociation, and thoughtless, unreflective physical action. His capacity to carry out the act – to fully realise his commitment – appears to depend less on holding to a conceptual proposition or logic and more on the grip of a corporeal state.

These qualities are then echoed as Brody makes his promise to Dana. Having repaired the device at considerable risk of discovery, Lewis plays Brody as agitated when he approaches Walden for a second time. His repeat attempt is presented in a way that resembles the first, but expressively amplified in recognition of the heightened force of will that is required to kill oneself not once but twice. Lewis stands beneath a ceiling light, and as he prepares to detonate the bomb he casts his gaze upward, eyes open wide and mouth frozen as his face glows in an evocation of glory achieved through martyrdom, the camera pushing in as the buzzing strings crescendo in anticipation. An insert shows his thumb tense as it readies to apply the final pressure. We hear Brody grunt with exertion, but there is a pause, a break between mental intention and bodily act – the hand stays frozen, a hesitation that allows a Secret Service agent to interrupt with Dana's phone call.

Damian Lewis's handling of Brody's exchange with Dana is a small tour de force. In contrast to the contained innerness that is elsewhere in the series such a mark of his performance as Brody, Lewis plays his halting responses to Dana in a way that charts the swift disintegration of whatever psychological fabric has until this point allowed the character to follow any purposive course of action at all. With each question and plea from Dana, there is a caught-breath pause from Lewis, his eyes darting from side to side like a trapped animal while he struggles to calculate a fitting answer; his face spasms, nerves misfiring; words come forth in fits and starts, as though they were the issue of nothing more than a reflex on the verge of stuttering breakdown; and when Dana pleads for her father to come home, Lewis produces a voice that sounds strangled from within, struggling to emerge, at once machine-like in its rapid robotic tempo but pitifully human in its helpless, panicked desperation. The closeness of the camera now works in concert with the actor's changed manner to evoke not concentrated, intent dissociation from the world, but a state of paralysed terror and disoriented mental spinning. It is in emerging from this condition that Brody is able to make credible his promise to Dana, but as performed by Lewis the moment is not one of reassuring sentiment for the viewer. 'Dad, you have to promise me,' Dana says. 'I need you. You know that.' As though in response to Dana's many repetitions of the word 'promise' and 'Dad', Lewis brings his performance to breaking point, his face shaking, chest heaving in silent sobs as tearful gasps escape with each tiny, crushed breath. But then there is a change. Lewis's brow relaxes and the strain runs from his face. His eyes glaze and settle on some distant but invisible point. Stillness overtakes him, at once within and without, as silence falls, his panicked breathing calmed. The greatest intensity of distress and confliction is in a passing instant replaced by an emptying absence of emotion (Figure 4.3).

It is in this newly opened void that Brody at last speaks with a measure of certainty: 'I'm coming home, Dana,' he says. 'I promise.' Lewis delivers the line with a dead calm, the distance in his eyes further colouring his cold vocal tone, as though Brody's speaking of this promise were akin to the rote incantation of a mantra. Lewis's performance of Brody's reversal thus brings to mind not the rediscovery of a submerged love, but the words of Abu Nazir that underpinned the suicidal murderousness of which he had proven himself capable only moments earlier: 'There is nothing left to think about. Clench your teeth. Say the Holy words. Remember Issa.'

Brody's pivotal reversal is aesthetically credible not because 'Marine One' supplies a plausible set of motives that can be ascribed to explain the character's actions. It is instead because previous moments of performance – most fundamentally the treatment of Damian Lewis in the pilot episode's dual mirror scenes – have prepared us for just such a sudden and seemingly

Figure 4.3   An emptying absence of emotion (*Homeland*, Showtime, 2011–)

inexplicable loss of what seemed to underpin a person's ultimate convictions. The significance that *Homeland* gives to the provisional in serial drama is thus an unsettling one. It is vividly captured in a storyline from episode eight ('Achilles' Heel', season one, episode eight) which forms a pre-echo for Brody's change of heart. In pursuit of Tom Walker, the FBI enlists his wife Helen (Afton Williamson) in a sting operation. Knowing Walker will call Helen's home to hear the voice of his ex-wife and son on their answering machine, the authorities trace the line and encourage Helen to entrap her former husband, saving him from Nazir's plot and protecting the nation from further attack. When Walker calls in the middle of the night, Helen manages to keep him on the phone, talking to him even though he does not respond. Eventually, as the trace is about to close in, Walker speaks. 'Helen,' he says. On hearing her long-lost husband speak her voice, she turns in horror towards Carrie. 'Oh baby,' she says into the phone, 'I've done a horrible thing.' As Carrie tries to seize the handset, Helen screams down the line: 'They're tracing this call – you've got to get out of there!' As in the case of Brody's call to Dana in 'Marine One', long-held commitments to another person and to a larger body of collective belonging are brought into irreconcilable conflict, and the switch of allegiance from one to the other is little more than the sound of a particular voice speaking a certain word. It is not a reassuring picture of love or loyalty, but more like an image of hypnotism and deep unknownness. What is gripping in these moments is our encounter with the mystery of what keeps us tied together, or so suddenly breaks our bonds.

Acknowledgements

Thank you to the editors for their patient and helpful feedback on an earlier version of this chapter. I am also deeply grateful to both Jason Jacobs and Murray Pomerance for their generous, incisive thoughts.

Notes

1   We only hear the words spoken by Carrie herself when she later recounts the event.
2   See Jacobs (2001: 435–37), and Jacobs and Peacock (2013a: 6–7).

3   Brody and Walker formed a two-man sniper team, Walker the shooter and Brody his spotter. This has more than practical importance to the plot – it is also thematically significant, evoking acute ocular sensitivity, a mortal bond of trust between soldiers, and the sniper's protective 'overwatch' role, which echoes Carrie's conception of her CIA duty.

4   Of course, in television drama we are always watching an actor filmed as they perform their character. My point is that the mirror scene makes this issue especially salient, difficult to put aside without overlooking a crucial aspect of the scene's significance.

5   See the interview with actor David Harewood in Hogan (2014).

Part 2

Television Performance and Collaboration

5

Approaching Performance in Contemporary *Coronation Street* (1960–)

James Zborowski

You are a television professional. The vast majority of your working time is devoted to the production of a single, long-running programme, which is broadcast several times each week. Your particular skill set includes the ability to act naturally in front of the camera and to make pre-planned interactions with others retain a feeling of freshness and spontaneity. You possess and perform a persona, which comprises principally a default attitude towards the world and others in it, sometimes underpinned by a familiar repertoire of gestures and phrases. It functions as a protocol for the encounters you undertake with others, and sometimes also as a protective armour. Others are aware of, and sometimes adapt to, this persona in their dealings with you; likewise, some of those you encounter have established personae of their own. You prepare carefully for your time in front of the camera, making sure you are well-informed about the interactions you will be participating in. The production set-up you work within, however, means that much of your preparation is either solitary, or in consultation or collaboration with others who will not appear on-screen with you. Typically, your on-camera interactions with others – that is, the things that are recorded and broadcast, the end product that the audience will actually see and hear – are not rehearsed collectively. Often, there is a little time, before recording, for a brief discussion, or a walk-through, to establish some general parameters, but then, without further ado, it is time to record.

The description above fits not only television news presenters and chat show hosts, but also soap opera actors. Recent research into the labour of television acting confirms what the number of hours of each soap opera broadcast per week, every week, already strongly suggests: the production schedules of modern British soap operas are very tight indeed. Moreover, the research demonstrates that 'little or no time [is] allowed for rehearsal prior to filming' (Hewett 2015: 75; see also Cantrell and Hogg 2016). This is true of much British television drama (and, one imagines, much television drama in other countries too), and it is especially so in soap opera, one of the most time-pressured production species of all.

What does this overlap between the demands of fictional and non-fictional broadcasting suggest? (Note that with a few minor modifications – mainly, replacing the word 'camera' with 'microphone' – the first paragraph could easily be made applicable to radio too.) I want to argue that it at least makes a plausible initial case that we might profit from adding to the toolkit we use to analyse performance in certain fictional television genres, including soap opera, concepts and approaches developed in relation to non-fiction broadcasting. The particular tools I will be recommending are ones that have been developed by Paddy Scannell.

If I may be permitted a few brief and almost certainly overly general statements to help very roughly stake out some terrain: when attending to much theatrical performance, some of the most rewarding analytical routes to meaning and value include (1) paying attention to the actor's prosody and line delivery (perhaps particularly when one has heard the same words delivered differently by many other actors before); (2) assessing the degree to which the performer has effaced or assimilated their self to the demands of this particular role; and (3) recognising the performance as a feat of coordination and choreography (sometimes literal) between performers who have, over many long hours, adjusted and dovetailed their individual performances into an aesthetic whole (or kept them deliberately and interestingly discordant). When attending to film performance, we will often do well to notice: (1) the integration of actors into a film's overall mise-en-scène, with a level of detail and care that often extends to re-lighting different shots in a single scene to sculpt star performers in light; (2) the endless possibilities of the relationship between the camera and particular characters, with the former able to bring to our attention the latter's smallest and most private gestures, to underline and participate in or to offset the character's movement and mood, and so on; and sometimes (3) the possibilities of ensemble staging to convey moods and relationships through characters' intricately choreographed, unfolding spatial orientations to one another and to the camera.

The areas of interest pointed to above are certainly not in total eclipse in television drama in general, or in British soap opera in particular. Soap opera has never lacked for actors who know how to deliver their lines to maximum effect. In the first episode of *Coronation Street* (Granada, 1960–2006; ITV, 2006–), Violet Carson, as Ena Sharples, through her pace and tone, packs disapproval, morbidity and acerbicness into her extended digression (humorously apropos of very little) in chat over the corner shop counter regarding choices of funeral songs. Few recent viewers of the same programme can have failed to have been delighted by the elaborate syntax and elevated register of Patti Clare's Mary and David Neilson's Roy Cropper. And only the seriously uncommitted viewer of contemporary soap opera will manage to avoid or overlook its periodic use of expressive or even downright flamboyant camerawork. This ranges from simple but effective moments of psychological expressivity, such as the camera singling out and describing a slow circle around Bethany Platt/Lucy Fallon to underline her isolation and distraction (episode 8880), to the use of mobile long takes to survey large ensembles, such as the 80-second shot that moves between and pauses upon different groups during a nightclub hen do (episode 8902). However, if our aim is to best do justice to British soap opera's most abiding and thoroughgoing characteristics and achievements, we may do better by beginning with different areas of interest. The three I would like to propose, all taken from the list of concepts that structure Scannell's account in *Radio, Television and Modern Life: A Phenomenological Approach* (1996), are *identity, authenticity* and *sociability*.

In preparing this chapter, I watched every episode of the programme from 8 April to 13 June 2016, then chose a sample of six episodes from within that range to form the basis of my close analysis (episode numbers are based on those provided by the 'Episodes' page on Corriepedia). I chose not to focus on a block of consecutive episodes in an attempt to control for any variance that might arise from a particular type of storyline dominating and/or reaching its climax, or the idiosyncrasies of a particular writer or director (I noticed that episodes are sometimes written and/or directed in blocks of around a week). Working with a small sample was a prerequisite for close analysis: it allowed me to watch each sample episode multiple times, with a range of different questions. The (admittedly elementary) attempt to 'control for variance' by choosing a spaced-out subsample of episodes from within the main sample was a way of trying to ensure that my analysis focused on what is typical of *Coronation Street*, rather than choosing particularly high-profile and/or dramatic and/or violent episodes and storylines, or 'stunt' episodes (such as the periodic live episodes), which may not be representative of the programme, and the performances within it, as a whole. My principal focus on my six-episode sample is supplemented with references to the history of *Coronation Street*, and to a much lesser extent other British soap operas. I am at present an occasional rather than an avid viewer of *Coronation Street*, but I viewed the programme religiously (in the company of my mother) in the late 1980s and early 1990s, and this particular viewing history informs parts of what follows. The ephemerality of individual soap opera episodes, and of some soap opera characters, means that some of my description below will no doubt make reference to characters forgotten or half-remembered by future readers. I cannot see a way around this, and in any case, readers with access to Box of Broadcasts will remain able to check the primary textual evidence for themselves. My more general references to well-remembered characters will, I hope, future-proof my contemporary snapshot to some degree. (For a brief formalist analysis of my six-episode sample which addresses matters beyond performance, see Zborowski 2016a.)

Identity

As Brand and Scannell note, careers that involve 'performing in public [...] may involve the projection of a carefully crafted identity and the management and maintenance of that identity in and through time' (1991: 203). Broadcasting is one such career, and although non-fictional broadcast performances (such as that of the DJ, Brand and Scannell's principal focus in their discussion) and fictional ones afford different possibilities for identity projection, both are part of what Langer terms television's 'personality system':

> Television personalities [...] become anchoring points within the internal world that each programme uniquely establishes in and for itself. They exist as more or less stable identities within the flow of events, situations or narratives which are presented in a particular programme at any given point in its cycle of repetition.
>
> (1981: 357)

The principal fictional genre Langer discusses is situation comedy – specifically, *All in the Family* (CBS, 1971–79), the humour of which, Langer argues, arises

> in part not from the element of surprise or novelty, but from anticipating when Archie will 'do something funny' the way he did it the week before. These ritualized routines come to represent his identifiable, reliable features as a 'personality'. Each episode one waits for Archie to call Mike a 'meathead', Edith a 'dingbat', get confused over large words, tell someone to get out of his chair, roll his eyes in exasperation or make a racist remark. These become the foibles to look forward to each week, […] which help to sustain a sense of familiarity and intimacy.
>
> (1981: 359)

Of course, soap opera's different structure means that, unlike sitcom, it does not in its episodes present self-contained narratives based on humorous misunderstandings or failed enterprises, and its *degree* of catchphrase repetition and stereotyping is less than sitcom's. Nevertheless, much of soap's 'familiarity and intimacy', along with other elements of its appeal, stem from its deployment of personalities.

Many of the most memorable and well-loved performers in British soap history are memorable and well loved not principally due to their ability to move audiences through their registering of emotional states in response to dramatic situations, nor perhaps (though this is an issue to be discussed further below) even due to their encounters and interactions with other performers, but rather due to their ability to project a slightly elevated, humorous, 'larger-than-life', often somewhat overbearing character portrait, a personality that announces and imposes itself on social situations, and has (except when these performers are called upon to lead a dramatic storyline) a one-size-fits-all response to those situations.

In the world of British soap opera, the tradition of larger-than-life personalities is perhaps especially pronounced in *Coronation Street*, which, as Richard Dyer (1981: 2–6) notes, emerges at the moment of the publication of Richard Hoggart's *The Uses of Literacy* (1957) and, in fact, places itself in self-conscious dialogue with that book and its nostalgic reverence for the 'full rich life' (Hoggart 1957: 132–66) of working-class culture, one element of which was conventions of performance (and sociability) emerging from the tradition of music hall performance and spectatorship (see also Lury 1995: 122–23). Indeed, some members of *Coronation Street*'s early cast, including Violet Carson, Doris Speed and Arthur Lowe, had careers in music hall and/or repertory theatre. Notable personalities from across the programme's run would include Ena Sharples (Violet Carson), Hilda Ogden (Jean Alexander), Bet Lynch (Julie Goodyear), Vera Duckworth (Liz Dawn), Reg Holdsworth (Ken Morley), Norris Cole (Malcolm Hebden) and Roy Cropper (David Neilson). In other soaps, we might point to the Dingles in *Emmerdale* (ITV, 1972–), or to Alfie Moon (Shane Richie) in *EastEnders* (BBC, 1985–). The viewer will be able to rely upon stock mannerisms, ways of talking, and ways of reacting from these characters. It is unusual for this to go as far as sitcom-style catchphrases, but not unprecedented. *Coronation Street*'s butcher Fred Elliott (John Savident) fell just short of a catchphrase in his verbal tic of saying many things twice, and punctuating the repetition with an 'I say!' When Bianca (Patsy Palmer) and Ricky (Sid Owen) returned to *EastEnders* in 2008, their reappearance was the subject of a BBC promotion in which Ricky stands outside The Queen Vic, and a scream of Bianca's famous, oft-parodied utterance 'Rickaaaay!' knocks him to the ground and shatters all of the pub windows.

Contemporary *Coronation Street* sustains a significant number, and proportion, of personalities in its cast. We have, for example, the amiably gormless Kirk (Andrew Whyment, who played a not-dissimilar role in sitcom *The Royle Family* (BBC, 1998–2012)), the avuncular and cheeky Lloyd (Craig Charles, who imports aspects of his persona from his long career as a television and radio personality), and the smooth-talking Dev (Jimmi Harkishin, who increasingly appears to be channelling the spirit of Harry H. Corbett as Steptoe Jr), as well as the already-mentioned Mary, Norris and Roy, among many others. We might note that both Whyment and Charles joined the cast of *Coronation Street* following successful roles in sitcoms (Charles played Dave Lister in *Red Dwarf* (BBC, 1988–99)). And just as *Coronation Street*'s early performers imported some of the register of the music hall into the programme, these performers bring a slightly declarative, slightly heightened style that makes a useful contribution to its overall texture.

A good example of a contemporary *Coronation Street* personality is Sean Tully, played by Antony Cotton, who has received several awards for his performance and is a favourite among fans of the programme. Sean's two jobs, as a machinist at the Street's underwear factory and a barman at The Rovers Return, mean that the character is dramatically placed to contribute to several interactions in addition to the ones arising from his own personal life. Let's look in a bit more detail at three scenes featuring Sean/Cotton, as a way of exploring further what the use of a series of personalities might offer to *Coronation Street*'s audiences and its production team.

These three scenes all come from the episode broadcast on Wednesday 4 May 2016 at 7.30 pm (episode 8896), and together comprise all of Sean/Cotton's appearances in that episode. The first scene Sean appears in is the sixth in the episode, and takes place in the café (Roy's Rolls). Sean is sitting with his boyfriend Billy (Daniel Brocklebank), a vicar. A slightly downcast Billy refers to being 'half-asleep' during morning prayer. Sean counters with enthusiasm: 'Ooh I was up with the lark! I could've come and cheered you on.' He leans towards Billy, offering smiles and raised eyebrows in between sentences. Sean tells Billy, momentarily raising his forefinger to emphasise the declaration, that he holds the cure to his tiredness. He then makes Billy close his eyes while he retrieves on his phone a holiday booking he has made for them. Billy is frustrated by Sean's committing him to time and expense without consultation, and thus begins the drama of the scene. This same structure and set-up is reprised later in the episode, in the next scene featuring Sean, when Billy visits his house. We join the scene just as Sean is complaining that 'This Ryvita's sucking all the moisture out of my mouth.' He is eating it with the aim of losing two kilos for the holiday. As Billy tries to get in with his objections, Sean is all enthusiasm and plans, jumping in a series of directions – his new shirt (bought for the holiday), the question of whether they should go all-inclusive, and the issue of whether Billy has been to the travel agent to pay the deposit. The final scene featuring Sean takes place in the Rovers. The villainous Pat Phelan (Connor McIntyre) enters with his building colleague Jason (Ryan Thomas) and orders 'Vitamin L'. Seizing on the opportunity for playful banter, Sean, indicating via his voice and a head-flourish that the conversation is not on a serious footing, responds, 'Limoncello?!' This is greeted with a stony face by Pat. Sean's smile drops. He clears his throat, looks down, and says, in a subdued voice, 'I'll get you a lager then shall I?' (Figure 5.1).

There are dramatic purposes to these scenes, but what I want to highlight is that a lot of the pleasure offered by Sean's presence in them is not much to do with the programme's ongoing plots. Rather, Sean is appealing

Figure 5.1 Sean (Antony Cotton) injects occasion into the everyday during his chat with boyfriend Billy (Daniel Brocklebank) (*Coronation Street*, Granada, 1960–2006; ITV, 2006–)

in the way that, say, chat show hosts are appealing. He is 'up', and his initial response even to contrasting moods like despondency is to try to bring its holder 'up' with him. He introduces occasion and enthusiasm into situations where others would be content to remain merely businesslike or matter-of-fact. He *performs* his personality. There is a self-conscious self-possession to the way he tries to draw others into his mood. He uses his gestures and his gait for effect. They are part of his way of being-in-the-world. In his discussion of 'Televised chat and the synthetic personality', Andrew Tolson (1991: 180) argues that two important components of 'chat' as a mode of speech and interaction are 'displays of wit' and the 'opening up [of] the possibility of transgression' – both of which are especially evident in Sean's abortive invitation to Pat to create such chat together.[2]

Sean, we might say, is good craic. His company is of the type that would be worth seeking out if one is after relaxed sociability, and the possibility of lifting a potentially humdrum experience. Broadcasting, as Scannell has pointed out, is a purveyor of sociability. It is so certainly as much as, and probably more than, it is a purveyor of drama and fiction. Scannell goes so far as to describe sociability as 'the most fundamental characteristic of broadcasting's communicative ethos' (1996: 23). Sean is a fictional character, but he is also 'a character' – a lively contributor to social interactions. The importance of this extra-dramatic, extra-narrative, para-social appeal of the performances of many British soap actors probably cannot be overestimated.

From a production perspective, there are also advantages to having a cast of personalities. Many of the highlights of script and dialogue in *Coronation Street* in recent years have arisen less from particular storylines than from particular characters. It is clear that writers take relish in providing certain performers with good material (to the extent that one might be forgiven for mistakenly believing that the writers of Roy and Mary's lines are twice as good, or have twice as much time, as the rest of the team). On-set, the bringing of well-delineated personalities by performers to their minimally or un-rehearsed recorded interactions with other cast members is a valuable extension of the working practice whereby actors arrive on-set with clear ideas about how their character would respond to this or that situation (Cantrell and Hogg 2016: 285–86). Moreover, in the case of such personalities, it will not only be the individual performer who knows how they are likely to play a moment; other cast members will also be better able to predict the performance choices and the projected persona of that individual. It is not a criticism of Cotton's performance or range to observe that the self-conscious self-possession with which Sean approaches social interactions levels out some of the differences we might, in other situations, expect to see in a character's dealings with different characters, and with intimates versus acquaintances. And this is far from being exclusive to Sean/Cotton. Indeed, one of the defining features of a strong personality, in the realm of soap opera and beyond, is that it places all manner of social interactions on its own, rather specific and unvarying, terms.

Authenticity

As well as including performers who project strong, well-defined personalities, contemporary *Coronation Street* also includes performers who are good at listening and responding to others. Tim is a chiefly comic character, played by Joe Duttine, who is very good at communicating an effort to decipher the reactions of the person he is talking to – usually, to see if his latest moment of comic subterfuge (such as, in episode 8920, trying to hide from his stepdaughter the fact that he has created a micro-brewery in his conservatory) or attempt to dig himself out of an awkward moment created by his own social blunder (such as, in episode 8887, backtracking when he realises that his friend's wife did not know that her husband was coming home from Spain to attend her surprise birthday party) is working. Perhaps more surprisingly, Jack P. Shepherd, who plays the scowling, blunt-speaking David Platt, is also a performer who does a particularly good job of appearing to react authentically to other characters. In episode 8903, there is a series of scenes in the Rovers backyard featuring David and his half-brother Nick (Ben Price). Nick announces that he will not be marrying his fiancée, because he lied about getting the all-clear following his head injury and now his symptoms are returning. A combination of script and performance generates the authenticity in this sample exchange:

Nick:	I had the scan today. Haven't got the results yet.
David:	Well you told me you had the all-clear, why'd you lie to me?
Nick:	Because I don't want her to know.
David:	Carla?
Nick:	Yeah well she's a mess. She needs someone who's stable, you know, a rock. It's a lose-lose situation.
David:	How'd you work that out?
Nick (increasingly impatient):	OK. I tell her, she leaves me. I don't tell her, I hurt her, she leaves me. It's a bit rich innit?
David:	What is?
Nick:	This. This pretend concern.

Given that the issue of sincerity is raised diegetically and dramatically, it is perhaps especially important that this scene achieves authenticity and sincerity. Authenticity is aided by the script. The exchange, without sacrificing clarity (especially for a regular viewer), remains somewhat elliptical and cryptic. Nick, in his agitated self-absorption, does not make himself entirely clear, and David has to press him for more information. Sincerity is delivered, in no small part, by Shepherd's decisions about how much silence to leave after his co-performer finishes speaking. His 'Carla?' follows hot on the heels of Nick's previous utterance because it is really the only available inference. 'How'd you work that out?' follows a slightly longer pause. In both cases, the length of the pause conveys listening and interest: in the first, a miniscule pause displays a desire to press the conversation on; in the second, the longer pause indicates that David is trying to figure out what

Nick is saying. David, here and in many other exchanges, is a surly and often a dismissive listener, but also often an attentive one (Figure 5.2).

In the world of conversation, seconds, even microseconds, really can make the difference between effects such as stiltedness, sincerity, calculation, caution, courtesy, wholeheartedness and rapport. Some insights that Scannell draws from a discussion of conversation analysis include: 'To hesitate is to *communicate* hesitation and that, in turn, is likely to preface a negative response of some kind' (2007: 185); and 'If you are going to accept an offer, do so immediately and without qualification. If you are going to refuse an offer, do so with apologies and explanations' (2007: 188). In the scene analysed above, there is great skill in the way that Shepherd, under rather restrictive production conditions, uses his judgement as a performer to approximate the revealing rhythms of everyday speech.

Is it naïve to use tools developed in relation to real-life talk to analyse talk in soap opera, where, as in all drama, we surely expect stylisation and, unavoidably, a degree of artifice? Perhaps. But to return to the points with which I began: we can make a case that, because of the way soap opera is produced, soap interactions are at least pushed closer to the realm of ordinary talk and the standards we use (often intuitively and pre-theoretically) to analyse the mood being created. Soap scenes are typically filmed in continuity, with even less rehearsal than other types of television drama. The aesthetic and

Figure 5.2 Nick (Ben Price) talks, David (Jack P. Shepherd) listens (*Coronation Street*, Granada, 1960–2006; ITV, 2006–)

production priorities of soap opera will funnel our attention more exclusively towards the interactions, and especially the *talk*, of its characters/performers than will other dramatic genres where mise-en-scène, or camera draftsmanship, or choreography command a greater share. As with the (I have tried to argue) related genre of non-fiction broadcast talk, there is an invitation, if we wish to accept it, to become connoisseurs of awkwardness and stiltedness (and its opposite, the successful pulling off of an interaction). Scannell has devoted attention on several occasions to failed sociability and failed rapport (Brand and Scannell 1991; Scannell 1996: 37–48; Scannell 2013: 128–52); it is related to what Karen Lury (1995) describes elsewhere as 'corpsing'.

I want to tentatively suggest that we can place most soap scenes into one of five categories. I want to further suggest that the stakes of failed authenticity in the delivery of these scenes differ from category to category, as does the likelihood of such failure, given how soap opera is produced.

First, there are scenes that revolve around *concealment*, in which a character or group of characters knows something that another character or group does not. There are scenes of *consultation* or *collusion*, where one character approaches another for advice about the situation they are in. There are scenes of *sociability*, where characters enjoy one another's company and offer mutual support. There are scenes of *confrontation*, often the result of a concealment coming to an end. And finally, and perhaps most rarely, there are scenes of *catharsis*. Embarrassment or tragedy has struck, and one character offers counsel to another. (In recent years *Coronation Street* has featured several excellent scenes of catharsis between Roy/David Neilson and Carla/Alison King.) The following generalisations about how soaps in general, and *Coronation Street* in particular, handle these scenes are not watertight, but I hope they are suggestive.

Given that most humans understand and thrive upon gossip, and the protocols and mental mind-reading that surround unequally distributed social information, scenes of concealment possess the virtue of being both instantly legible and engaging for most. However, this same acute audience awareness, drawn from personal experience, means that such scenes are hard to pull off with entire conviction, and a measure of artifice has to be accepted. For example, at the beginning of episode 8903, Nick receives, and rejects, in quick succession, two mobile phone calls from his consultant concerning the return of his brain injury symptoms. He fobs off his fiancée, Carla, with a weak cover story about the caller being a wine merchant who does not know that Nick is no longer a restaurateur. The scene ends with

Carla, alone in shot, looking suspicious. The problem is neither that Ben Price fails to deliver a performance of an unconvincing lie, nor that Alison King fails to communicate the closing note of suspicion. It is, rather, that the authenticity of the encounter is significantly dented by the fact that neither party appears to be reacting fully and authentically to the other's behaviour through their interaction. Each individual character's behaviours and reactions, the viewer may well feel, is too obvious not to be probed further by the other party. This may partly be an issue of pacing and duration. In many respects, contemporary soap opera moves too fast (see Zborowski 2016b), but here full dramatic credibility may be being sacrificed to a structure which requires concealments to fester over and across scenes and episodes. These problems are resolved more satisfactorily in instances where we are not dealing with two essentially decent characters who each have their reasons, but with a duplicitous and manipulative character who *is* good at lying, and who only lets the mask drop in private communion with the camera. A recent example of such a character in *Coronation Street* is Pat Phelan, who is regularly granted such moments to underline, and allow the audience to revel in, his villainy.

Scenes of confrontation also risk coming off as stilted, though for different reasons. A good example comes in episode 8880. In a scene at the Platts', eight characters are gathered, and have an argument. On one side, there is the Platt/Tilsley family: Audrey, Gail, Nick, David, Sarah-Louise, Bethany. On the other, there is Lauren, a girl who has been bullying Bethany, and Lauren's mother. Earlier in the day/episode, Nick has thrown Lauren's bag to the ground, destroying in the process a bottle of shampoo. The conversation quickly becomes heated. Recriminations and then insults are exchanged. However, until David gets up to eject the two unwanted visitors, all characters remain rooted to the spot. They further oblige the demands of framing and cutting by meticulously observing a rule that only one character should speak at a time. The result is that the confrontation lacks heat, and therefore loses credibility. This example is particularly striking because the number of participants in the confrontation makes its rigidness and odd formality especially noticeable. However, similar problems beset other confrontations. In the viewing sample, the numerous confrontations between Michelle and Caz often feel similarly forced, because the performers stand and deliver rebukes that do not overlap. This is no judgement on the acting abilities of the cast. It is much more likely to arise from the production processes of soap opera. Performances of chaos, counter-intuitively, require more rehearsal than performances of orderliness, and perhaps they cannot be pre-scripted to the same degree as other scenes, or accommodated within the often cramped sets and the shooting technique that soap favours. It may also be the case that to get into the heightened emotional state demanded of confrontation is another achievement that the schedule and set-up of soap production constitutes extra barriers to. (It may also be a matter of different production cultures *between* soaps. I admit to speculating on a limited viewing sample here, but my impression is that *EastEnders* on average does confrontation better than *Coronation Street*.)

Sociability

Of the five categories I have proposed, we can see *concealment* and *confrontation* as forming a pair. In both, the characters are in some way at odds. They are not committed to a shared view of or approach to the situation they find themselves in. By contrast, the other three species of interaction – consultation/collusion, sociability and catharsis – precisely *are* ordered around the characters working together to create a shared mood and reality. The set-up of soap production and the skill sets of broadcasting professionals make it likely, I would suggest, that at least a minimum threshold of authenticity and sincerity will be achieved in such encounters. A commitment to a shared enterprise is what brings the real-life creators of *Coronation Street*, in front of and behind the camera, together in the first place. It is not naïve to suppose that, if the performers are also equipped with an ability to comport themselves naturally in front of a camera using scripted lines, this commitment and rapport can aid the pulling off of sociability on-screen. Indeed, some of the deepest pleasures of soap opera viewing are provided in the rapport that exists between particular pairs of performers. In the case of the strongest personalities, discussed earlier, the screen chemistry that exists is that other characters are simply catalysts for the reactions of those strong personalities. But in some cases, characters are genuinely reagents: they bring things out of one another that other characters do not. The late 1980s and early 1990s were a golden era of such relationships in *Coronation Street*. Key instances would be Reg Holdsworth and Curly Watts, Fred Elliott and his nephew Ashley, Jack and Vera Duckworth, Rita and Mavis, Derek and Mavis, and Alec and Bet. Interestingly, this period also constituted something of a drought for television comedy variety and double acts, which had appeared earlier in variety shows, and sitcoms. *Coronation Street* was, perhaps, filling this void in this period. More recently, the love and friendship between Beth and Kirk, Gemma and Chesney, Steve and Lloyd, Steve and Tim, and Tyrone and Fizz have provided their own forms of enjoyment. To watch these pairings is to watch characters and performers comfortable in and enjoying one another's company.

I hope in the above to have offered a slightly new angle on the analysis of British soap opera, or at least *Coronation Street*. Drama is driven by conflict, but broadcasting is driven by sociability. Soap opera partakes of both of these sources of engagement. If I have emphasised the sociable side of the equation, this is not because I underrate the importance or the achievements of soap's plots and dramatic structures (though I do believe these elements of soap have their limitations, which deserve to be explored, as I have done elsewhere). It is rather because I believe the interactions that are not based around confrontation and concealment to be among the greatest sources of value and pleasure offered by soap opera, even if it is confrontation and concealment that command the most attention in the way that soaps are marketed, summarised, and otherwise circulated. Calling attention to this value, and to the tools developed elsewhere in television studies that might help us to articulate it, might aid a fuller appreciation of hitherto critically neglected elements of soap opera, and a fuller account of what kind of a thing – and how remarkably complex and multifaceted a thing – soap opera is.

Notes

1 See https://coronationstreet.fandom.com/wiki/Category:Episodes (accessed 8 March 2019).
2 One of Tolson's main case studies in his chapter is Dame Edna Everage, a comic creation of Barry Humphries, and the features being discussed here suggest why sociable chat in broadcasting has been a niche that has been occupied, not exclusively, but successfully, repeatedly, and in a particular way, by camp performers. It is worth noting briefly that Cotton's potential as a chat show host was, it would seem, recognised within the industry, leading to Cotton hosting *The Antony Cotton Show* for ITV in 2007. However, the programme was not a success, and did not return after its first series.

6

Don't *Curb Your Enthusiasm*: Visible Bonhomie and the Ontology of Improvisational Comedy

Tom Brown

Improvisation has received relatively little attention within television and film studies. There are a number of reasons for this but, as Virginia Wright Wexman has noted:

> Though many films make extensive use of improvisation, it is usually difficult, if not impossible, to obtain accurate documentation about precisely when and how the improvisation occurs. Nonetheless, most critics would agree that our response to the acting in many scenes filmed by directors like Godard, Bertolucci, Scorsese, Rivette, and Altman is partly determined by our sense of a spontaneous, unrehearsed quality which these directors, in various ways, manage to convey. This *effect* is susceptible to analysis, even though its *causes* are often unclear (1980: 29; original emphases).

In addition to the practical reasons Wexman notes, I would speculate that the place of improvisation in Godard, Bertolucci et al. has been neglected because it sits in tension with the valuing of these *auteurs*' control over their works – to allow space for improvisation is to cede considerable control to the agency of the performer. My focus in this chapter, a television sitcom, is not so encumbered by restrictive notions of moving-image art. The difficulties and, for me, undesirability of focusing on production ahead of the finished text mean that, like Wexman, I examine the improvisational as an effect, as a *style* of performance. However, it is in the peculiar ontology of improvisational comedy performance (which is peculiar in the way it draws attention to its ontology – to a level of its 'realness') that the 'causes' are, to some extent, intrinsic to the analysis. Moreover, it is part and parcel of twenty-first century television and film fictions that the boundary between the inside and the outside of that text is more porous than in the work of the film auteurs Wexman cites, due to cross-textual and extra-textual marketing strategies. Like most UK viewers, I did not consume *Curb Your Enthusiasm* (HBO, 2000–) on broadcast television but mainly on DVD. DVD 'extra' material, which is much concerned with the work of the actors 'behind the scenes', will here be especially fruitful and appropriate material to draw upon.

Curb Your Enthusiasm centres on the social travails of 'Larry David', writer, comedian and co-creator of *Seinfeld* (NBC, 1989–98). The speech marks around 'Larry David' are necessary because Larry David plays a fictional-ised version of himself (I will call the *character* simply 'Larry' from here on; 'David' refers to the actor-writer-producer) in a narrative populated by other celebrities playing themselves (Ted Danson, Richard Yorke, Ben Stiller etc.) but also more fully fictional characters such as Larry's wife Cheryl (played by Cheryl Hines) and his friend and business manager Jeff Greene (played by Jeff Garlin who *is* a friend of David's but is not his manager).

Ethan Thompson, following Brett Mills (2004), labels *Curb* and other recent sitcoms in a similar stylistic vein, 'comedy verité'. The information he offers on the programme's production methods, as well as his caution about the limits of the 'improvisational' in 'hand-held improvisational comedy' is instructive:

> While [each] scene is improvised, two high definition video cameras shoot the action simultaneously, with one always on David because so much of the comedy depends on his reactions. There is no rehearsal of scenes, just one quick blocking, then the first take is a general master shot. The scenes will be improvised several times as the director and actors sharpen the scene and get plenty of coverage to ensure it can be cut together.
>
> (2007: 69)

As Thompson and others (Dolan 2006: 92) have noted, this shift to a greater emphasis on post-production takes the show away from traditional sitcom-making and closer to something documentary in nature. Of course, a documentary mode of production does not necessarily involve a greater proximity to the 'real' (many would say it doesn't even with more fully doc-umentary content) and Thompson is at pains to point to the considerable effort of construction that lies behind *Curb*'s ostensibly on-the-fly produc-tion method:

> According to [Robert] Weide [a collaborator of David's on the show], the scenes in *Curb* are restaged until all the necessary coverage is obtained and the dialogue and comedy refined and repeated during the process as well. From reading this description and by looking at the resulting scenes, it is clear that the 'hand-held, improvised' (observational) approach to the show is another type of coverage filmmaking. The cameras may be hand-held and the content improvised, but the editing of the shots follows classical Hollywood patterns.
>
> (2007: 70)[1]

One might also add to Thompson's final point that the fact of there being one camera always on David[2] means it has perhaps not travelled so far from traditional sitcom's 'three-headed monster' as have other shows such as *The Office* (BBC, 2001–03; see Mills 2005: 61). However, shifting from pro-duction causes to stylistic effects, the ultimate style of *Curb* allows the per-formances' roots in improvisation to be appreciated as a crucial part of their texture. What the show's creators *leave in*, and leave to be seen, are the actors' search for the right way forward in the scene/situation, awkward pauses and also the actors' self-evident amusement.

Curb Your Enthusiasm represents a radical peak in Adam Kotsko's dis-course on *Awkwardness* (2010), an extended essay that also takes in the British version of *The Office* and the film work of Judd Apatow. Kotsko has less to say about the improvisational style of performance than about the awkwardness of the show at a thematic and narrative level, but what he does say may be quoted in its near entirety:

> A […] source of awkwardness comes from the format of the show itself: David writes the overall stories to a certain degree of detail, but individual lines are improvised. In some cases, that leads to remarkable lines that probably wouldn't have arisen otherwise, as when Larry claims that his environmen-tal activist's wife's choice of toilet paper is 'like what you'd find on a whaling vessel' – disarming Jeff, who 'breaks character' with his laughing repetition of the line. Most of the time, however, the result is a strange dialogue style that seems somehow less 'realistic' than directly scripted dialogue would be. The

interaction is staged, and all the participants know it, but they can't fully inhabit the characters they are supposed to play (who are in most cases themselves!) as they would be able to if they had preset lines to deliver. Real life and fiction become blurred, to the point where my above claim that Jeff 'breaks character' in his spontaneous response to Larry's line becomes very difficult to adjudicate: what could breaking character possibly mean in this context? The characters are caught in a strange in-between space where their very lack of the normal apparatus of scripted acting calls even greater attention to the fact that they're acting.

(2010: 70)

There is some slippage between character and actor both in the logic of the argumentation and the use, at times, of first names for both (surely 'breaking character' is something the actor would do?). However, Kotsko offers a number of jumping-off points that I will use loosely to structure the remainder of this chapter. Kotsko's particular interests lead him to linger on the resonance between what he sees as the 'strange dialogue style' engendered by the improvisational approach and the theme of awkwardness. My concerns are more with the kinds of moments represented by the laughed response of Jeff/Greene to 'whaling vessel' (the complex relationship between character/actor being crucial in this and other such moments). These illustrate what I am calling the 'visible bonhomie' between the performers. This quality of joyful fellow feeling and connectedness is the virtual antithesis of awkwardness. It is a crucial pleasure of *Curb Your Enthusiasm* and of other contemporary television and film texts, and may indeed be a necessary, more throaty release from the nervous laughter associated with such 'cringe' comedy. An analysis of audience pleasure is clearly more speculative than close attention to textual effects, and a sufficiently broad appreciation of the performance style of *Curb* must take in the maintenance of dramatic tension (and awkwardness) in Cheryl's interactions with Larry and the latter's often greater freedom with his male co-stars.

'Less Realistic' (But More Real?)

Before devoting the discussion to the abandonment to bonhomie that *Curb* sometimes, perhaps increasingly, allows itself, I want to consider a passage of performance work that is more characteristic of the baseline for the show's particular style. A dense relay of useful examples comes approximately 20 minutes into the second episode of season three, 'The Benadryl Brownie'.

The girlfriend of 'Richard Lewis' (who within the show and outside is Larry David's close friend) is determined that he will take her to the upcoming Emmy awards. However, she is suffering from the effects of a severe allergic reaction to peanuts (Richard suggests a dress for the awards show that matched her face would be 'made of turnips and blood') but, as a Christian Scientist, she will not take any medication. The Benadryl brownie in the episode's title indicates the plan to feed her a popular antihistamine without her knowledge. The passage I am examining begins with the end of a scene where Larry is trying to convince Susie Greene (Susie Essman) to give him her brownie recipe. Unfortunately for Larry, hers is a secret recipe handed down through the generations of her family. In an attempt to create a reciprocity of divulged secrets, Larry tells her 'a secret nobody knows, not even Cheryl [...] I might be losing a testicle [...] It's not definite'. It is clear to the viewer, at least from narrative context, that it's an invention. Susie seems to believe it but it makes little difference. She closes the door on him refusing to countenance giving him the recipe. We cut from a medium shot of Susie moving back over the threshold to a reverse angle of Larry, shifting his weight from one foot to another rapidly and shouting indignantly with rising intonation (Figure 6.1).

Larry slightly tilts his head and raises his eyebrows as he shouts conveying, perhaps, sincere disbelief that such a personal 'revelation' would be met with a seemingly petty refusal *and* the calculation of his attempt to project disbelief in order to maintain his front/perhaps change Susie's mind. Shifting more firmly to the ontological agent behind this performance (the actor-writer-producer), David allows the hint of a smile to colour his expression: there is an unmistakeable twinkle in those eyes, a quiver in the shape of his lips as he closes his mouth at the end of his sentence and a tightness around his mouth that suggests suppressed mirth. It's not hard to imagine that, in the shooting of this scene, David might have 'corpsed'.

Corpsing (sometimes known as 'breaking') is under-theorised in film and television studies.[3] 'Corpsing' means the actor's inappropriate laughter during the performance of a scene – the term derives from the fact that the worst time for involuntary laughter would be when one is performing being a corpse. The visible proximity of the performance to corpsing, or to something like it, is important to recent 'comedy verité' and to the dominant,

Figure 6.1 'I just told you I was going to lose a ball!' (*Curb Your Enthusiasm*, HBO, 2000–)

'improvisational' strains of mainstream American film comedy over recent years (I take Apatow to be the pre-eminent figure in this). Strictly speaking, there is no corpsing in *Curb Your Enthusiasm*, though it is present 'extra-textually' on a 'gag reel' on a solitary season of the show's DVDs (season six) and its frequency during shooting is commented upon in numerous DVD bonus features. Unlike the popular British sitcom *Mrs Brown's Boys* (BBC, 2011–), the machinery of production never intrudes into *Curb*'s fictional world and there is no acknowledgement of the fact that it is a TV show. The regular corpsing in *Mrs Brown's Boys*, like in famous television comedies before it,[4] is partly explained (though, in this recent show, now in a much more 'meta' way) by the filming in front of a live studio audience. *Curb*, in contrast, is exclusively shot on location and possesses no laugh track. While there is no corpsing per se, there is often what one might call more broadly 'authentic laughter' in David's exchanges with his co-stars and celebrity cameos – a crucial facet of the visible bonhomie I will focus on more fully in the later stages of this chapter. While the DVDs' extra-textual material makes it clear David is regularly breaking down into laughter, this is not always left in, at least not in scenes of anger, frustration and awkwardness, of which there are many. Indeed, on more than one DVD extra feature, we are told that David is always corpsing, particularly when another actor, working in character, is shouting at him. In a feature called 'The History of Curb … Even Further' (season five DVD), Susie Essman claims, 'We shoot a million takes … and especially when I shout at Larry. He gets hysterical laughing.' In the same segment, Ted Danson wryly notes, 'This is not a highly trained actor we are dealing with.' One might then wish to evaluate David's performance within the context of the potential expressivity of 'inexpert acting', or to understand the performance style of *Curb Your Enthusiasm* as a response to his limitations as a performer. There is some merit to the latter view especially. However, contrary to Danson's comment, I would want to stress David's remarkable skill and *expertise* as a performer. A brief example from season two, episode six ('The Acupuncturist') can serve as an example.

For reasons too convoluted to explain, Larry takes it upon himself at a party to convince Mr Weiner (Ed Asner) to give his son Barry (Jeremy Kramer) some money. In a typical move of failed social dexterity, Larry relays a story about how, when he was a struggling comedy writer in New York, his father decided to give him some of his inheritance early, which provided the opportunity to pursue his vocation and ultimately succeed with *Seinfeld*. Larry tells Mr Weiner, 'I was working as a dry cleaner [Weiner's son is a failed comedy writer currently working in a deli]. I didn't tell him. I had no money … but I needed to support myself…' David runs the lines together more rapidly than appears natural and raises his intonation on the 'him', 'money' and 'myself', conveying a sense of slightly over-eager entreaty and a rush to hit certain 'beats' (Larry, like David, is of course a writer). His head and face (on-screen for the duration of 'I didn't tell him' to 'myself') bobs while he maintains quite insistent eye contact with Weiner. During 'I needed to support myself', David deploys his face and eyes especially skilfully, looking at Weiner, looking down and forwards very briefly, closing his eyes, then looking at Weiner again very rapidly and raising his eyebrows oddly high and opening his eyes oddly wide, which puts strange emphasis on the end of 'myself' (Figure 6.2).

An enormous challenge to describe or reproduce in prose form due to the extreme rapidity of facial shifts (and shiftiness), David expertly communicates the clumsy 'tell' through Larry's rapid glance at Weiner. The effort

Figure 6.2 David's purposeful deployment of Larry's unconscious 'tell' (*Curb Your Enthusiasm*, HBO, 2000–)

of both concealment (concealing the machinations of his work on Barry's behalf) and checking Weiner is on board is visible to the audience but sufficiently credible within the broadly 'naturalistic' tenor of the exchange; we can see that Mr Weiner may well see Larry's designs (indeed, he promptly disinherits his son after the exchange, believing Barry has put Larry up to this) but the speed and pseudo-subtlety of Larry/David's vocal and facial appeals to paternal love and duty give *some* justification to Larry's self-satisfied smile after Weiner has gone off to find himself a probate lawyer. In this moment, I am minded of a number of formulations in Alex Clayton's theorisation of the standard functioning of comedic performance. For example: 'The friction between "meaning to" and "not meaning to" is a comic strategy I want to call *incongruous intentions* […]. The clash of opposing qualities is permitted only by recognition of comic design' (2012: 52). David the actor means to display his clumsiness but Larry the character does not. The skilful *balancing* of this incongruity is a marker of David's conventional comedic skill. Returning to the passage from 'The Benadryl Brownie' episode, the suppressed smirk as Susie closes the door on Larry is a different kind of 'tell', where the actor's relationship to character subtly changes. Here a refinement of vocabulary is needed. I've suggested a familial relationship to 'corpsing' but that it is like but not quite like this – not quite corpsing because that involves putting the fiction to one side, corpsing being only fully and truly visible to us in 'live' productions or extra-narrative, extra-textual material such as 'bloopers'. One could more accurately designate what David is doing as 'suppressed corpsing'. However, this is a description of the technicalities of performance and does not indicate the significance at the level of the fiction. I would suggest 'narrative corpsing' as my tentative term for what the actual episodes of *Curb Your Enthusiasm* do. It is not ideal but better than the obvious alternatives: 'fictional corpsing' would imply that the corpsing itself is fictional, when the point is it is a 'real' element poking through the fiction; using the more common American term for corpsing, 'breaking', would lead us in a different direction if one called it 'narrative breaking'. 'Narrative corpsing' suffices because the point is that the corpsing remains, partially at least, *within* narrative in the fact that, I would contend, it enriches or at least textures characterisation rather than interrupting it in the manner of, say, Peter Cook trying to make Dudley Moore break down.

In contrast to David, Cheryl Hines does not allow herself/is not allowed to intrude upon her representation of Cheryl David in quite this way, though the machinations of the show's improvisational practice certainly colour her performance. After Susie has shut the door on Larry, via a brief establishing shot of Larry and Cheryl's home we cut to Cheryl who has clearly been told of the Benadryl brownie plan. In this nine-second-long shot, Cheryl is relatively slow to look up and at Richard and Larry (it takes half the shot's duration), formulating her thoughts to herself as she haltingly relays her understanding: 'OK, so, um … you wanna … put some Benadryl in some brownies.' There are notable pauses recreated in my transcript but also a slight halt between 'Bena' and 'dryl'. Her gaze remains cast downwards until 'wanna' and she purses her lips tightly before the look up in a move that, in combination with subsequent raised eyebrows and head tilts on 'put some Benadryl in some brownies', conveys Cheryl's thought process as containing the question, 'Am I to take this seriously?' Hines' eyes are large and wide, a frequent occurrence in her embodiment of Cheryl's marriage to Larry. Here, an observation of the text (Cheryl/Hines' eyes are often wide in 'surprise') can be enriched by 'extra-textual' knowledge. On the same DVD featurette cited earlier, to behind-the-scenes footage of the shooting of her frequent performance of shock, Hines says:

> Larry [David] wouldn't even show me the story outlines for the first two seasons. That's why I'm always saying in the episode, 'What?', 'What are you talking about?', 'What did you do?', 'I can't believe you just did that' because I'm hearing it for the first time. ('The History of Curb…Even Further' (season five DVD))

'The Benadryl Brownie' may be season three but such is the particular ontology of *Curb*'s performance style, enabled if not demanded by its production method, even for a performer given the story outlines (with any scene comprising only a paragraph of outline anyway) the 'search for the appropriate response' is frequently visible. In Hines' case, the stock phrases she repeatedly reaches for ('What did you do?', 'I can't believe you just did that' etc.) and, in the case of the scene examined above, the lengthy pauses, hands in front of her mouth at the start of the shot and the delay in looking upwards, can be seen as a performative 'holding pattern', giving the improviser time to formulate the most effective way forward. Of course, the nine-second shot analysed here might be a record of the hundredth time Hines responded to lines said to her by Lewis and/or David)[5] but the crucial point is that the 'search for the appropriate response' is left in. More than simply a typical baseline of Hines' performance in the series, such examples of the moment before a response are, I would contend, a defining characteristic of the performance style of *Curb Your Enthusiasm*.

Here, it is worth returning to the earlier-cited passage from Kotsko:

> Most of the time […] the result is a strange dialogue style that seems somehow less 'realistic' than directly scripted dialogue would be. The interaction is staged, and all the participants know it, but they can't fully inhabit the characters they are supposed to play […] as they would be able to if they had preset lines to deliver.
>
> (2010: 70)

Kotsko makes some evaluative leaps (that the actors 'can't fully inhabit the characters') but it is worth dwelling on where his thoughts seem to abut with a key concept in performance analysis: naturalism. The words of the script become, within broadly naturalist fiction, the spontaneous expressions of the character's interiority. Kotsko's scare quotes aside, I think it would be more logical for him to have written that *Curb*'s 'strange dialogue style' is 'somehow less *realist*' than 'directly scripted dialogue'. We are accustomed to understanding 'realism' and the 'realist' as a set of conventions that vary over time and in different national and production contexts. The broad realism that Kotsko seems to be drawing upon contains naturalistic acting which, by definition, attempts to hide the fact of its being acting. *Curb* even pokes fun at this notion in season seven, episode ten ('Seinfeld') with the launch party for the book written by Jason Alexander (playing a pompous, simple-minded version of himself), *Acting without Acting*. Jason explains the title thus: 'You don't want to see the actor at work. You wanna hide the technique. It's acting, it's doing the job, it's the craft, so you hide the actor's effort.'

'Naturalist' acting is a broad category within television and film and encompasses relatively ostensive traditions such as theatrically trained Great British Actors and more post-'method' approaches. In the latter, the words of the pre-existing script are absorbed into an intensely psychologised and 'rounded' notion of character. Comedic performance operates very differently and, as I am arguing in this chapter, improvisational comedy performance is more particular still. Cheryl Hines is in fact the performer in *Curb* who remains most consistently within a broadly naturalistic register (comedic ostensiveness is not her primary role in the ensemble). However, the visibility of the 'search for the appropriate response' means that the delivery of dialogue has a different impact to conventional naturalism and thus perhaps explains Kotkso's claim that the style is less 'realistic'. Might one not, however, counter that the moment examined above is actually more 'real'? Of course, in a real-world exchange we do not know what other people are going to say to us and, therefore, the techniques of improvisation enable a potentially more authentic, and therefore more real, response. This seems to be the aim in approaches to improvisational filming we understand more readily as 'naturalist', such as Ken Loach's (see, for example, Johnston 2001: 6). However, what perhaps prevents *Curb*'s style being received as more real (or realist) for Kotsko is a certain intensity, a dialling up of the performer's look to their fellow actors and the self-reflexivity of the text as a whole, a self-reflexivity that is familiar to comedy: 'By its very nature, comedy undermines our involvement with the characters, barely maintaining a dramatic illusion [...] [and] invites us to observe plot machinery *as* machinery' (Naremore 1990: 114).

Cheryl resists Larry and Richard's attempt to enlist her help in baking the Benadryl brownies throughout the earlier-cited scene. Larry tries various arguments and seems finally to succeed once he has made the point that the allergic reaction occurred 'in our house'. Cheryl looks down at the brownie mix they have brought her, furrows her brow and quietly relents, 'OK.' Richard/Lewis's response remains within character and at the same time can be read as an index of the progress of the improvisational exchange between the three actors. Richard responds to the tenuousness of Cheryl's assent by backing out the door, raising both hands up, nodding rapidly and muttering the incomplete sentence, 'I think we should... you know...'. His motion is made in an effort to draw Larry out of the room, his actions all communicating the character's sense that the men should 'quit while they are ahead' and leave before Cheryl changes her mind. Reading this through our sense of the way the scenes are put together and improvised (a reading encouraged by textual elements embedded and well-established through the series I would suggest), Lewis's tentative and sensitive exit communicates the actor's awareness of having found an appropriate end point – there is not a script that would have given this exchange a preset end. The greater apparent looseness and, arguably, 'realness' of the flow of inter-character exchanges is felt also in Larry letting Richard leave, telling him, 'I just want to talk to Cheryl for a second'; it is characteristic of Larry that he rarely knows when to quit while he is ahead. Larry talks to Cheryl in a more intimate register trying to assert that the Benadryl plan is 'not really a big deal'. As he is withdrawing, 'Can you bake this stuff? I don't even know... Can you bake? I don't know. Can you? I've never seen you bake anything, I'm sorry!' With the latter words, as he backs out the door, Larry starts to chuckle. Cheryl flashes him a warning look as he leaves the room. The genuineness of David's smile as he offers brief, nervous laughter fits the narrative situation of a husband who is constantly pushing his luck with his long-suffering wife.

Performance analysis confronts issues of both genre and gender as one notes the difference in Larry/David's genuine laughter in this interaction with Cheryl from those with his male co-stars. It is not 'visible bonhomie', not because the laughter is nervous but because it is not shared. It is Cheryl's or, rather, Hines' role, to stay firmly 'in character' as is evident in the furrowed brow as she glares at Larry's laughing exit. It would be unfair to suggest that the programme does not afford space at other times for bonhomie between the couple. However, the more free-flowing encounters with other characters (male and female) are anchored by this sitcom's situation of its lead character as an overgrown man-child married to a dutiful and civilising wife, which is nothing if not conventional within sitcom and the culture at large. This framework sits in tension with the show's deserved reputation for 'pushing the boundaries', including, for example, its clever invocation of paranoia about Los Angeles' lesbian and black 'communities' ('The Bowtie', season five, episode two) and the agency, space and force afforded the performers Rosie O'Donnell and, more recurrently, Wanda Sykes. However, if one wanted to drill down further into the politics of its performative pleasures, one would have to note that the roles of these two actors are largely antagonistic, and the performers with whom the viewer might see more of the indulgence in visible bonhomie (with Jeff Greene often and Richard Lewis sometimes, with Bob Einstein and, as examined below, Jerry Seinfeld) are male.

'What Could Breaking Character Possibly Mean?'

The use of the term 'indulgence' above leads to an important issue, as the risk that *Curb* runs, especially given its narrative setting among wealthy Los Angeles celebrities often playing themselves, is of being accused of solipsism. Visible bonhomie between the performers, particularly those 'playing themselves' is, or undoubtedly can be, a source of pleasure. However, the potential audience or reviewer response to a comedy, 'They clearly had a good time making it,' might be at least as much a reaction to the perceived self-indulgence of the enterprise than an expression of enjoyment at the rapport of the on-screen performers, especially if one feels excluded from the on-screen community (male, wealthy, largely white) having such fun. The reliance upon improvisation certainly enables spontaneity and a certain kind of 'authenticity' but needs to be balanced with tight story cohesion if the pitfall of narrative formlessness is to be avoided. I would argue that *Curb* has navigated this terrain successfully and consistently for eight seasons because of the obvious emphasis David and his collaborators have put on story. Jason Alexander, who appears prominently in two seasons, states, 'Larry's version of "unscripted" is the Magna Carta. It is so *written*' (emphasis in original speech; 'The History of Curb … Even Further'). Alexander is responding to a situation that, reportedly, has seen David's story outlines get much longer as the seasons of *Curb* have progressed (ibid.). While the individual scenes are created through the collaborative work of improvisational performance, the overall trajectory in which they fit follows a pattern written by David, and there are often strong threads that structure individual episodes (awkward and risqué themes such as 'affirmative action', which gives the title to an episode in season one, and cunnilingus, central to season three's 'Krazee-Eyez Killa'); that develop across whole seasons (the focus on the production of Mel Brooks' Broadway show *The Producers* in season four, for example); and that even recur between seasons (a joke about the potentially injurious impact of cunnilingus from 'Krazee-Eyez Killa' recurs in episode eight of season seven where it plays a narrative role).

There has, however, been a noticeably greater and fuller 'indulgence' in visible bonhomie as *Curb Your Enthusiasm* has progressed. In season eight, episode five, for example, in the opening scene when Larry joshingly mocks her accent and way of speaking, Susie/Essman is shown to us, briefly, on the edge of 'corpsing', something hard to imagine in earlier seasons (key to her characterisation and narrative role is that she is the most constantly aggressive and antagonistic character towards Larry). I would see season seven as a turning point in an increasing abandonment to narrative corpsing, chiefly with the appearance of Jerry Seinfeld.[6] David and Seinfeld worked closely together for the best part of a decade on their earlier sitcom and their particular rapport in both personal and professional endeavours is evident on-screen, underpinned by the specific compatibility of their performance styles which both combine the untrained and improvisational. The major season seven narrative thread is *Curb*'s backstage presentation of a *Seinfeld* reunion show and episode six features much interaction between the pair as they share an office for the writing and production. For example, as Larry and Jerry drive to lunch, they discuss Larry's managing to make their assistant quit, something Jerry admits was his plan when he insisted Larry was the one to speak to her: 'I thought, this Bulvan will walk in there and, in two minutes, she'll be out in a huff.' Jerry smiles very widely and looks at Larry at the end of his sentence and, throughout, is on the edge of laughter. However, Larry's reaction is what interests me more, particularly at the word 'Bulvan' where David (for now it seems unmistakably the actor 'losing it') throws his head back, reaches over and grabs Seinfeld, hits him playfully and collapses into strangled spasms of laughter (Figure 6.3).

Figure 6.3   David's utter abandonment to laughter and bonhomie (*Curb Your Enthusiasm*, HBO, 2000–)

This is the moment across all the seasons where Larry/David most fully gives himself over to laughter or, rather, it is the moment we see this most clearly outside of the DVDs' 'extra', more straightforwardly documentary material.[7] The shared laughter, the evident pleasure Jerry/Seinfeld takes in making Larry/David laugh and the latter's utter abandonment to this laughter is a marker of their visible kinship. Larry's grabbing of Jerry is a kind of tactile, masculine bonhomie nowhere else evident in David's *Curb* performances, and the use of the word 'Bulvan' (a Yiddish word I had to look up, meaning a 'loud mouth know-it-all; a boorish, brutish person') underlines a fraternal bond expressed in shared cultural and comedic heritage (i.e. New York Jewish). Perhaps some viewers may feel alienated, remote from such a personal connection but I would contend, however anecdotally, that the affectivity and contagion of laughter draws us in. I called the bonhomie of the moment 'masculine' because I take this kind of playful hitting to be, stereotypically at least, more characteristic of male camaraderie and friendship and perhaps also of more juvenile iterations of such relationships (e.g. the man-child trope in contemporary comedy). However, I should at this point stress that I don't believe there is anything inevitably or innately 'masculine' about the improvisational performance style I am examining. It is equally visible between the female performers of *Bridesmaids* (dir. Paul Feig, 2011) and, in a TV sitcom context, between Ashley Jensen and Ricky Gervais in the latter's *Extras* (BBC/HBO, 2005–07). Indeed, in *Extras*, visible bonhomie is a crucial facet of the interactions of Gervais and Jensen, and its absence from his awkward on-screen interactions with Stephen Merchant are more defining of the latter fictional relationship.

Within *Curb*'s diegesis, the 'Bulvan' exchange is the moment most unmistakably akin to 'corpsing' that we see from Larry/David. Like Kotsko, we can see this as 'breaking character' but could go on to ask, 'What could breaking character possibly mean in this context?' As noted earlier, 'corpsing' is defined as inappropriate laughter by the actor, inappropriate to the construction of a fictional character. The idea presumably goes that if an actor starts laughing uncontrollably, unable to say their lines, hit story beats or marks, we are made aware of the presence of the lines, beats and marks in the first place and taken out of the fiction. The appropriateness of uncontrolled laughter by the actor in a scene where they are joking around with their fictional and (often in the case of *Curb*, real) friend clearly has to be judged differently. Something repeated more than once by David and his collaborators in the DVD extra material is the authenticity of the laughter on the show and the illogicality of the absence of diegetic laughter on regular sitcoms. In 'Roundtable Discussion with Larry and the Cast, Recorded Live at New York's 92nd Street Y' (season eight DVD), we have the following exchange:

David:	The one thing that I enjoy about the show is that when people laugh on the show, it's genuine laughter because they're hearing something for the first time and they're reacting honestly and spontaneously to it, so the laughs on the show are real […] in the first couple of seasons, I'd be in the editing room and people would say, 'You can't leave that laugh in, it looks fake.' But it's a real laugh! […] So I just say, 'Well, we're leaving it in.'
Jeff Greene:	I think we might be the first show to do that … because so often when you're watching a sitcom someone says something unbelievably funny and everyone just says their next line. It's, like, crazy! Did you hear what he just said?!
David:	Right. Everyone is so clever all the time on these shows, everybody saying funny things constantly but nobody's laughing!

For a proper theorisation of 'realism', 'verisimilitude' or 'naturalism', it would be necessary to unpick at length the reasons that a technician or collaborator might describe a real laugh as 'fake' in the editing suite. Suffice to say for now that David is making a case for the value of the ontological realness of the visible markers of pleasure and connectedness (particularly laughter) of *Curb*'s on-screen performers (the interviewer in the above exchange then segues the conversation naturally into a discussion of the casting of David's friends on the show). Even if the show sees lines and exchanges developed through improvisation repeated so many times that one might almost see the process as simply a rehearsal leading to a finished text, the genuine laughter is clearly often left in (it is hard to imagine that the Larry David–Jerry Seinfeld moment cited above was anything other than the first time they did that particular thing on camera).

The method utilised for making *Curb* is highly relevant to this discussion but it is not essential to an evaluation of its effects. Indeed, in line with viewing the process as filmed rehearsal, it is conceivable that the performances could be more practised so as to 'work out the kinks' and create a more fully realised 'natural' surface – this would ultimately help 'hide the actor's effort' as per the words delivered by 'Jason Alexander'. A 'kink' or 'crinkle' in the performance is an apt metaphor for what we actually have, because the very presence of these is what characterises the performance mode of *Curb*; the actor is no blank canvas for the delivery of character but protrudes upwards and out through the surface of the character in a way a kink or a crinkle might in an otherwise smooth sheet or fabric. This metaphor of performative protuberance is subtly different from standard notions of 'ostensiveness' (see Naremore 1990) and complements and extends Clayton's notion of the 'comic twinkle' (2012). Clayton shows how comedies frequently rely upon our recognition of comedic intention and its frequent co-presence with the character's fictional obliviousness. The suppression of mirth visible in Larry/David's mouth as Susie closes the door on him fits with this but I feel there is a peculiar pleasure in observing the actor's *effortful* suppression of mirth. One might speculate that this draws us into, makes us collude even more intensely with, the play-acting that gives Clayton his title. The peculiar ontology of comedy, and the more particular ontology of improvisational comedy, also necessarily engages the spectator with something 'real' the performers and filmmakers draw our attention to. Future analyses might helpfully focus discussion on qualities such as inter-performer 'chemistry', 'rapport', even individual 'charisma' as specific textual elements, notoriously intangible if not vague and woolly notions. My contention is that close attention to the minutiae of performance and to the concrete work of the performer is essential if one is to properly theorise the function of important qualities such as these.

Notes

1 Further information about the editing of the show and the particular challenge of editing its overlapping dialogue is contained in Dolan (2006: 92).
2 This was enabled by an increase in budget from season two onwards (Dolan 2006: 92).
3 Karen Lury's 1995 essay is the often-cited piece on the phenomenon of corpsing in television.
4 Peter Cook and Dudley Moore as 'Pete' and 'Dud' in *Not Only But Also* (BBC, 1965) is probably the most famous instance of corpsing in British comedy history.
5 The average ratio of filmed to used material is 30 to 1 according to Dolan (2006: 91–92).
6 Seinfeld is also seen briefly in the finale of season four.
7 The moment also very clearly resembles the non-fictional format of Jerry Seinfeld's web-based *Comedians in Cars Getting Coffee* (Crackle, Netflix, 2012–), which it predates.

7

Tears, Tantrums and Television Performance

Amy Holdsworth and Karen Lury

This chapter focuses on the contradictory and ontologically confusing aspects of performance in reality television. This kind of television is messy and confusing because it makes quite considerable and often contradictory demands of both performers and audiences. Reality formats require viewers to evaluate the success – tied to the performance skill – of cast members who operate across the epistemological boundaries that would normally allow audiences to distinguish between scripted and unscripted conversations and events. Furthermore, in a manner that is now very familiar to the television audience, reality programmes feature 'real people' who are, at the same time, paid participants ('cast members') whose 'actual' lives are voluntarily constrained and managed. Most audience members are well aware of the fact that reality formats are intentionally designed to manufacture an apparent intimacy with performers and to generate high levels of drama or jeopardy. Performers who are successful in these contexts therefore commonly offer both fabricated and actual expressions of extreme emotional states. As a consequence, television viewers can experience quite dramatic shifts in their perception and judgement as to the authenticity of performers in different reality show contexts. Performers may be perceived as either genuine or fake (or both) and the audience's emotional response to them can swerve – in a matter of seconds – from affection and empathy to disbelief and disdain.

Crying and tears, we argue, are endemic to reality television and in this chapter we use examples of these emotional displays to highlight the significant 'slipperiness' of television performance. Cultural theories and histories of crying foreground the ambiguity of tears: tears resist coherent interpretation but as they leak from the body, appearing as material evidence of or witness to an internal emotional state, they demand a reaction. Tears are provocative: not only do they leave open frames of performance to question and contestation but they *invite* speculation. What we wish to capture is how crying and the presence (or, indeed, absence) of tears crystallise key questions for the study of television performance and open up notions of authenticity, judgement, evaluation and the competency of the performer. Our focus, in this chapter, is to explore how reality formats encourage audiences, through the repertoire of tears, to employ different kinds of judgement. We argue that viewers judge the apparent sincerity of the performer (or, in effect, the 'performance skill' of the contestant/cast member to appear sincere) and that this is done in the context of an ongoing judgement as to the ethics or intentions of the programme maker (which determines our reading of specific kinds of 'performance cues' established within the programme's narrative). In our own evaluation as to the significance of tears, crying and performance, we conclude by demonstrating how, within what is, in effect, a moral economy, television supplies an emotional education, a context in which the recognition and value of sincerity, empathy and compassion may be exposed, debated and contested.

If understood as a genuinely unplanned physiological and psychological response to events, tears would seem to offer a clear example of actual emotion and thereby a prompt to the audience to respond to the subject with empathy or pity. For Pansy Duncan they present a 'somatic vocabulary marked by both authenticity and legibility' (2011: 180). Yet Duncan is writing in the context of the (relatively) coherent vocabulary of tears within the openly fictional genre of Hollywood melodrama, whereas within the realm of reality television the distinction between fact and fiction is much more blurry, and tears are subject to judgement and evaluation. Too many tears suggest artifice and excess, while crying without tears is often read as the ultimate signifier of inauthenticity. For example, in the summer edition of 2015's *Celebrity Big Brother* (Channel 5, 2011–), with only a few days to go until the final, veteran television presenter and news anchor Eamonn Holmes enters the house to host a show within the show – a final quiz where the remaining housemates are subject to the judgements of the audience. Having been voted the most 'fake' housemate by the public and accused of engaging in a 'showmance', *The X Factor* (ITV, 2004–; Fox 2011–13) alumnus Chloe Jasmine and her on/off screen partner Stevi Ritchie are gently interrogated by Holmes. Leaning over his desk and overtly mediating between the housemate and the viewing public, he states: 'Chloe Jasmine, we have to work out what the public are seeing as fake between you two and let me just say one word: tears.'

Here Holmes exposes the conflict between a general understanding that 'crying' is often a managed performance but where, nonetheless, the perceived absence of 'real' tears may also condemn the housemate/performer as inauthentic. Despite the machinations of the production team and the continual, yet often hollow, pleas from contestants caught up in the emotional turmoil that it is 'just a game', *Big Brother* is typical of the reality show format that frequently values 'realness', 'honesty' and 'sincerity'. Chloe Jasmine's theatrical response to Holmes acknowledges, therefore, her failure to *perform authenticity* through crying 'adequately'; she claims: 'Tears are words that the soul cannot express. I'm sorry it hasn't been believable but to me it's been very real.'

The term 'crocodile tears' refers to the insincere expression of sorrow: 'It stems from the ancient belief that crocodiles, in order to lure their prey, would weep. The unsuspecting prey would come close, only to be caught and rapidly devoured, again with a show of tears' (Dent 2009: 66).[1] While crocodiles are thought to weep while they eat, a result of the physiological structures of their breathing, humans are largely understood to be the only species to cry as an emotional response. In Tom Lutz's historical study of 'cultures of tears' he acknowledges that there are a 'great variety of kinds and causes' of crying (1999: 21). Yet one of the 'perennial strands of the cultural history of crying' is the distinction between 'good' and 'bad' tears, and where 'those that are not "genuine" have been held in contempt' (1999: 31).[2] These are the tears of the crocodile – artificial, manipulative, dangerous even – a 'breach of not just etiquette but of ethics' (1999: 21) and running counter to a twentieth-century belief in the 'naturalness of tears' (1999: 33). Tears, then, operate across a strangely porous dividing line – between what seems to be natural (and authentic) and that which may also be perfectly and wickedly faked (and therefore deceptive). For Annette Hill, this tension is exhibited in the 'twin issues of performance and authenticity [which] frames discussion about the authenticity of visual evidence in popular factual television' (2005a: 449).

Scholarship on reality television has often emphasised the role of judgement as at the heart of the audience's engagement with the genre. Hill's work, for instance, considers how audiences are 'engaged in the critical viewing of the attitudes and behaviour of ordinary people in the programs, and the ideas and practices of the producers' (2005a: 453). Citing John Ellis, she argues that

> audiences of reality programming are involved in exactly the type of debates about cultural and social values that critics note are missing from the programmes themselves: 'on the radio, in the press, in everyday conversation, people argue the toss over "are these people typical?" and "are these our values?"'
>
> (2005b: 9)

Challenging conservative fears regarding the blurring of reality and fiction, Helen Piper has also argued that reality formats such as *Wife Swap* (Channel 4, 2003–09) should not be read

> so much as an infiltration of 'real life' by drama and performance, but a desire to dramatize, to actively compose and engage in sense-making narratives about the 'real lives' of ourselves and others, and to apply narrative moral logic to an artificially sequestered and unaccountable private space.
>
> (2004: 286)

The relationship between private lives and public discourse is revealed in Piper's account of the opposition between normality and difference constructed in *Wife Swap* and how it offers, reiterating Ellis and Hill, the audience a way of 're-moralising' life decisions.[3]

Tears, crying and the ethical implications of both, we argue, are central to this moral economy. As visible evidence they offer a distinctive materiality to performance within reality formats since their presence or absence help establish the terms of the judgement used by audiences, through which they evaluate the skill, integrity and appeal of particular individuals. Whether as part of the 'performing' or the 'authentic' self, tears invite interpretation and speculation. The multiple meaning of tears allows them to operate in many different ways: as expressions of pain, joy, grief, mirth, anger, frustration, triumph or pity. They often 'resist interpretation' while 'demand[ing] a reaction' (Lutz 1999: 19). As a form of communication, Lutz argues, 'the meaning of tears is rarely pure and never simple and thus no simple translation of the language of tears is ever possible' (1999: 25).

In her study, *Having a Good Cry*, Robyn Warhol outlines several different historical understandings of crying. Firstly there is an 'expressive' model, in which tears are read as an 'outward sign of internal emotional states', and secondly a 'performative' one, whereby the body's affective response brings 'into being the emotional states they betoken' (2003: 15). Analytical models of naturalistic modes of performance in fictional film and television would seem to draw upon the first understanding as they seek out the ways in which the body of the performer is able to express the internal state of the character. In this context, tears are central to the body's communicative repertoire. These 'drop[s] of limpid fluid secreted by the lachrymal glands' (*Oxford English Dictionary*) are seen to offer material evidence for performer and audience of the relationship between internal state and external sign, and the presence of emotion. Balancing precariously on the line between internal and external and notions of surface and depth (we speak of crying, for example, as welling up), tears and crying prompt a kind of ontological uncertainty between 'real' expression and what we understand as 'faked' performance.

A clear example of this confusion emerges in season six (episode eight) of the reality format and modelling competition *America's Next Top Model* (UPN, 2003–06; The CW, 2006–15; VH1, 2016–). In this sequence, the models are being set up for a photoshoot in which menthol sticks are used to provoke tears – so the audience are witness to a physiological prompt employed to fabricate distress. As the models submit to the clearly irritating treatment, with several 'crying' and wincing in pain throughout the shoot, one model, Nnenna, breaks down: it seems, as Tyra Banks (their mentor and host) suggests, that 'fake' tears have provoked 'real' crying. At first it might seem as if the audience is being cued here to respond with sympathy to Nnenna. Previously, the episode has built up to this moment of release as earlier scenes have foregrounded Nnenna's struggle with her relationship outside of the competition via the confessional features of the series format – phone calls back home, 'backstage' interviews and voice-overs and a 'girls' talk' with supermodel Janice Dickinson on the competing demands of work and family. However, for a long-term audience who are familiar with its format, the response is likely to be more complicated. Knowing that the series will be dependent on creating jeopardy and emotional responses from its cast, the audience will be anticipating displays of extreme and dramatic antics from the models. This means that they may be as confused as Tyra is as to where, during Nnenna's crying, one form of tears begins (fake/'performative') and the other ends (actual/'expressive'). In this sequence the performative and the expressive merge, or at least overlap, leaving the host and the audience uncertain about how they should respond (Figure 7.1).

In this episode's 'challenge' the models' brief is to make crying artful and beautiful – trails of mascaraed tears running down cheeks, brows lightly furrowed and eyes full of feeling. Within the framework of the photoshoot, their task is to communicate a depth of feeling within the photographs: to think sad, without losing the light, framing, composure or composition. The performance skill required from the models is somehow to communicate authentic emotion and, potentially, to do so by drawing on personal and genuine unhappy experiences and emotions. While doing this they are also required to manage their bodies and expression so that their crying is appreciated as beautiful, rather than an embarrassing and possibly snotty mess of bodily fluids. In addition, the sequence acts almost as a meta-critical reflection on the hypocritical aspects of the programme and the industry the

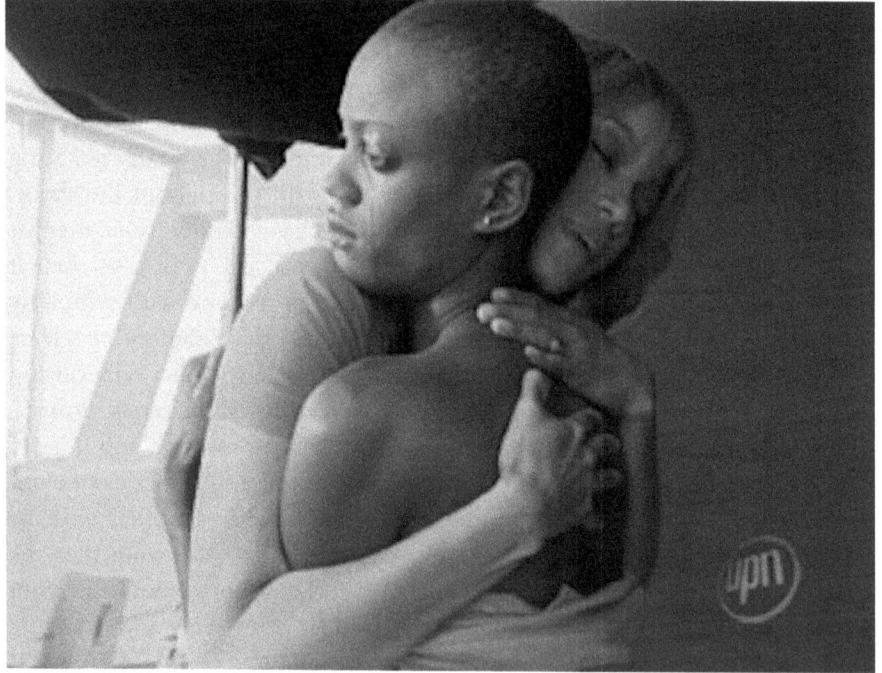

Figure 7.1 Tyra comforts Nnenna (*America's Next Top Model*, UPN, 2003–06; The CW, 2006–15; VH1, 2016–)

models are seeking to enter, since the competition, and the demands of the industry, insist that the models must also cultivate, manage and perform a persona that is necessarily contradictory and inconsistent – models are asked to be driven, ambitious, humble, respectful, resilient, sincere, charismatic and submissive (Figure 7.2).

In this instance, the 'cathartic' value of Nnenna's tears is insisted upon by Tyra and Nnenna herself, and this bout of crying is seen to allow her to get it out of her system so she can move on with the competition with renewed ambition and self-discipline.[4] Here then, Nnenna's tears are not simply about self-expression and a cue for the audience to sympathise with her as a character but an overt opportunity for Nnenna as a competitor within a fabricated format to 'get back in the game'. Employing a Foucauldian lens, Duncan's study of the film melodrama reminds us that the solicitation and regulation of tears forms part of a gendered disciplinary culture within which, other scholars have argued, reality television has established

Figure 7.2   The result of Nnenna's photoshoot (*America's Next Top Model*, UPN, 2003–06; The CW, 2006–15; VH1, 2016–)

a central role. As Wood, Skeggs and Thumin remark, the modes of judgement the format invites 're-routes and re-embeds classed and gendered [and racial] distinctions through a conservative ethics of individualization and self-improvement' (2009: 140).

In line with this observation, in an infamous judging and elimination from cycle four of *America's Next Top Model*, Tiffany Richardson, a young African American woman, fails to perform the appropriate levels of tears and emotion when she is ejected alongside another contestant from the competition. While her peer, Rebecca, breaks down into floods of tears, Tiffany smiles and makes light of the situation. Tyra praises Rebecca for her suitable response ('Rebecca, I admire your emotion right now. It shows to me that this was something that's very important to you'), then admonishes Tiffany for not taking the situation seriously. Tiffany immediately interrupts 'Looks can be deceiving', she says: 'I'm hurt. … I can't change it, Tyra. … I'm sick of crying about stuff that I cannot change. I'm sick of being disappointed.' Tiffany's failure to quietly submit to her mentor's lecture and her refusal to conform to a narrative of self-improvement through self-determination unleashes Tyra's wrath in what has now become a notorious and uncharacteristic display of anger – 'I was rooting for you!' she screams at the young woman, 'We were all rooting for you!'

In essence, Tyra's tantrum reverses the emotional conventions of the format. The inadequate display of emotion (real or faked) from her fellow performer causes Tyra's own mask to slip and the 'authentic' self of the normally 'caring' host is instead revealed. In other contexts this revelation might damage the standing of the presenter, yet reality television is based around the cultivation of transformation and the logic of the reveal. Tyra's anger, eyes glaring, brow contorted and flame-red hair shaking, epitomises those

> raw and spontaneous outbursts of emotion, what Laura Grindstaff (2002) refers to in relation to the talk show as the money shot: those moments when there is an eruption of anger, a breakdown of tears or a poignant moment of self-revelation.
>
> (Wood, Skeggs and Thumin 2009: 139)

In this respect, the possibly accidental revelation that Tyra may not actually be the saintly 'sister' figure she appears to be is not a problem, since the audience is far more invested in the speculation, contestation, judgement and evaluation of such moments. When one of the 'main viewing pleasures lies in detecting the moments of ontological integrity when people are not "acting for the cameras" but are apparently "being true to themselves"' (ibid.), Tyra's outburst simply provides the audience with the 'raw' material that the reality show promises.[5] Tears are similarly meant to offer readable and material instances during which the performer's self-control breaks down and the real self can be glimpsed; for example, the infamous tears of hard-nosed news anchor Jeremy Paxman in family history format *Who Do You Think You Are?* (BBC, 2004–) purportedly reveal an unexpected emotional openness. Yet the ambiguity of tears – scripted and unscripted, beautiful or gross, actual or faked – makes them open to question, however rampant they may be within television's wider 'commercialization of feeling' (ibid.).

Perhaps the most significant text in television's culture of tears has been the singing competition *The X Factor* (ITV, 2004–; Fox, 2011–13). The format deploys this commercialisation of feeling through a variety of familiar ways, manipulating emotion both narratively (the infamous sob stories) and formally (the long pauses and percussive heartbeats of the soundtrack). In a controversial moment from the 2011 US version of the show, 13-year-old contestant Rachel Crow stands in the bottom two with fellow competitor Marcus Canty. Not wishing to personally end the dreams of either of the hopefuls, lachrymose judge and former pop starlet Nicole Scherzinger, after a lengthy period of indecision, decides to take the vote to 'deadlock' (in the result of a tie the television audience vote determines which act gets sent home). 'Performing' distress, while the shouts and alliances of the studio audience can be heard in the background, she dabs at her eyes with a tissue, holds her hand to her brow and clutches at her heart: a series of gestures not unfamiliar from a theatrical repertoire of melodrama. Rachel, praised elsewhere for the 'maturity' of her voice and performance, continues in this mode and offers the judge her advice: 'Please don't cry, it's OK. I'm good with anything.' 'Performing' resilience, the cracks in her voice both belie her emotional state but also signal her control. As host Steve Jones explains the rules to the audience at home he is flanked by the two hopefuls. The studio audience chanting her name, Rachel sighs and smiles sweetly while Marcus nervously twitches on the spot. Cutting between the three shot and close-ups of the contestants, the grammar of the scene, familiar from a range of formats, emphasises suspense and builds towards the 'money shot', in Grindstaff's terms. The reveal – that Rachel has actually come last in the public vote – eventually comes, and given the clear preferences of the studio audience it is entirely unexpected by the young girl and apparently by the judges. When her name is called, her eyes widen in surprise and as Jones congratulates Marcus, Rachel falls to her knees, sobbing loudly and wildly – her chest heaving and husky wail drowning out the host's attempts to get on with the show. Head judge Simon Cowell rushes to the stage and Rachel jumps up to embrace him before her mother also hurries on to comfort her. We cut between the action on the stage and the hysteria which has now also consumed Nicole Scherzinger who weeps into her hands as she is consoled by fellow judge, Paula Abdul. As the studio director cuts into a close-up of Rachel and her mother among the huddle of 'concerned' judges and producers we hear and see the daughter seek reassurance from her mother: 'You promised mommy! You promise it's OK?' Yet despite the intensity of the scene, as her mother cups Rachel's face in her palms, no clear tears are visible.

What can we make of this moment? We have shared this scene on numerous occasions with colleagues and students and it nearly always prompts a deeply ambivalent response. We are also aware that in describing the scene our own choice of language will guide the empathies and judgements of the reader; for example, when we suggest that Rachel smiles 'sweetly' others may feel that it is more appropriate to say 'smugly'. Ambivalence might be generated by the cynicism engineered by a familiarity with those 'predictable patterns of feeling' (Warhol 2003: xvii) constructed by the programme, and as we have suggested, and studies of reality television audiences have illuminated, viewers' responses often bounce between trust and suspicion (Hill 2005a: 459) and their 'sympathies and pleasures are invoked simultaneously alongside derisions and judgments' (Wood, Skeggs and Thumin 2009: 136). The multiple and simultaneous responses to this scene are also bound up with cultural constructions of and ethics surrounding the fact that here the contestant is a child performer: there is *Schadenfreude* at work in the precocious child humbled by her defeat,[6] but the spectacle is also troubling as a 13-year-old girl is 'humiliated' on a public stage. Based on our perception and interpretation of her grief but the apparent absence of tears, judgements arise relating both to her failure to 'perform authenticity' and condemnation of the producers' exploitation of a child participant.

Rachel Crow's elimination offers yet another example of the 'money shot', the viral moment, the instant when the mask falls away. Here is the child finally acting like one –we could alternatively recognise Rachel's exhibition as a genuine tantrum, uncivilised rage, or the horrifying destruction of innocence, or does it simply peel back to reveal yet another layer of performance? The spectacular 'failure' of Rachel here demonstrates the layered frames of performance required by the talent format, and the complex set of skills required by performers as they must manage the often conflicting demands of such formats as both talent competitions *and* popularity contests. By becoming television personalities, reality show performers are required to inhabit key values of authenticity, sincerity and sociability (Bonner 2011). Yet the notion of an emotional 'reveal' promised by *The X Factor* and other reality formats is predicated on the problematic belief in an 'authentic core self' that could be revealed. However, in our ambivalent response we actually demonstrate the impossibility of identifying this core self with any certainty or extrapolating the 'authentic' inner self from a 'performed' exterior, however 'real' the situation attempts to become. Perhaps the tension here is less between performance and authenticity than the performance of scripted and unscripted (or improvised) emotion. We might question whether, within the context of the emergence and popularity of forms of scripted or structured reality television (*The Hills* (MTV, 2006–10); *The Only Way is Essex* (ITV2, 2010–); *Made in Chelsea* (E4, 2011–)), unscripted emotion has become harder to spot and the 'reality' in 'reality shows' is less and less likely to convince. The distinctions between 'backstage' and 'frontstage', to employ Erving Goffman's influential terms (1969), have become increasingly indivisible in these soap operas structured around the lives of their participants and, like tears, reality television performance is increasingly hard to anatomise.

The art historian James Elkins similarly suggests in his book, *Pictures and Tears*, that despite the physical materiality of tears: 'There is no way to tell honest tears from deceptive tears, and there is no way to anatomize a tear, and tell which part is which.' (2001:20) Elkins also suggests that 'crying is so common it might mean nothing, or just about anything' (2001: 19).[7] Yet he still wishes to assert, as we do, that 'crying' means *something* however murky its origin, and however suspicious our response may be to the appearance and authenticity of the tears that fall. Our interrogation of the meaning of tears and the performance of crying in reality formats is both compelled and complicated by our belief that across contemporary television – in news, documentaries, award ceremonies and charity telethons as well as in 'serious' drama – people seem to cry *all the time*. Tears on television are no longer confined to the beauty pageant, Oscars ceremony, or soap opera. The pervasiveness of instances of crying on television underlines Elkins' concept that tears represent 'nothing, or just about anything'.

So what is the value of our attempts to untangle the meaning of tears within these modes of television performance? To repeat Lutz's argument, tears may often resist interpretation but they demand a reaction, whether that is snorts of derision, 'studied inattention' or 'gestures of comfort or sympathy' (1999: 21). As such, tears are an evocative and provocative part of the communicative realm of television. Despite their murkiness, Elkins suggests that 'tears do one thing that separates them forever from the inarticulate parts of our inner life: they leak from our eyes, and run down our cheeks. They show, without room for doubt that something has happened. They are witnesses' (2001: 29). On television – and perhaps in everyday life – tears also serve to make others witness to another's unhappiness or joy and ask for a response. Even when we cry alone, the production of tears, the materiality of its wetness, your snotty nose and damp tissues, points to the fact that while crying is one of the ways we make ourselves visible to ourselves, tears would also seem to be produced always 'as if' others could see us. In a sense, then, television and tears similarly defer to an implied, wider audience even when that audience isn't watching.

In their sociality, tears and crying could play a key part in the potential of television to inform our reading of performance within what Lauren Berlant has called the 'intimate public sphere' (1997: 4). This potential is reliant on television's broadcasting capacity, its continuing domesticity and its pervasive presence, its familiarity, its accessibility and its ordinariness, as well as its didactic and pedagogical role. Tears and crying, we would suggest, are where and how performance on television supplies an emotional education, a context in which judgements about how we recognise and value sincerity, empathy and compassion can be exposed, debated and contested.

What is important for our argument is not that the crying seems justified, well performed, moving or authentic. It is more that tears provide evidence of the ultimately inarticulate and embodied nature of each individual's distress, anger or joy and that crying both on and before the television screen is a ubiquitous, everyday occurrence. The very unreliability, the open-endedness, the 'nothing and anything-ness' of tears means that they reveal that something cannot be said but there is still 'something' that is being expressed.

So? That's all very well, but isn't it dangerous to mix emotion with politics and civic society? Couldn't we use this argument to defend the increasingly toxic combination of emotion in politics that is seemingly exemplified by the Twitter tantrums of the 45th US president? Well, one answer would be that it is precisely the success of emotional responses and the way in which the rhetoric of the far right is legitimated via emotion that means we have to take crying seriously. And in wanting to validate the revelation, the performance and evidence of emotion in public life – in fact something that Berlant herself is ambivalent about – we can rely on the insights of other television scholars such as Misha Kavka (2008) and Kristyn Gorton (2009) as well as philosophers such as Martha Nussbaum. Indeed Nussbaum suggests in rather grandiose terms that: 'Walt Whitman was correct, I think, in his suggestion that there is a "public poetry" about the emotions that can be made the basis for the public culture of a pluralistic democracy' (2003: 402–03).

What we are suggesting is not simply that it is nice to have a 'good cry', or that it is good for news anchors, politicians and reality show performers to surprise us with their emotional competence and openness. It is rather that such displays offer us opportunities to exercise judgement about why people cry and to experience its contagious qualities: it is about recognising and acknowledging emotion as routine and as ubiquitous. Crucially, it demonstrates that emotion as an embodied, conscious and irrational relay system informs not just our 'private lives' but our public selves. Just as importantly, this should not be about 'dealing', therapeutically or in a Foucauldian (disciplinary) sense 'with' our own or others' emotional responses, akin perhaps to the concept of 'finding closure' or reinforcing the cathartic model of tears, but rather allowing for the fact that tears and crying are always open to question. As Elkins claims – 'tears are not like clues in a murder mystery, where everything is revealed at the end' (2001: 28).

This understanding of the value of tears and emotion brings us back to early assessments of television as a 'cultural forum' (Newcomb and Hirsch, 1983), and here we draw upon one final example of television tears taken from the hit Channel 4 series *Gogglebox* (2013–). In each weekly instalment the viewer at home is invited to go behind closed doors into living rooms across the country to watch carefully selected kinship groups watch and respond to highlights of the past seven days of television. The conceit of *Gogglebox* is arguably the rehearsal or performance of the democratic promise of television in a broadcast era and the dream of its continued vitality.

In the sequence in question the *Gogglebox* families are watching an episode of another Channel 4 show, *Educating Yorkshire* (2013), a fixed-rig reality series documenting a year in the life of a state comprehensive school in West Yorkshire. The episode in question focuses on a student, Musharaf, who is preparing for his English GCSE oral exam with dedicated teacher Mr Burton. Musharaf's preparation is complicated by the fact that he has a severe stammer. We do not have the space here to offer a fuller account of this remarkable episode but, instead, we wish to consider the emotional relay system set in motion by the series and its reiteration through *Gogglebox*. In the final assembly for the school leavers, Musharaf, the boy who had been bullied and didn't have a voice, takes to the stage to deliver a warm and self-deprecating speech. His fluid delivery, aided by speech therapy techniques of tapping and listening to music on headphones, and his heartfelt gratitude to the teaching staff has much of the graduating class and faculty in tears and was widely celebrated after its broadcast as 'moving the nation'. Screened for the *Gogglebox* families, this tearful consensus is reinforced. With some initially expressing intolerance for Musharaf's stammer, his triumphant speech and shots of fellow male students openly crying are intercut with close-ups of the *Gogglebox* cast's reactions: watery eyes, lumps in throats, wide smiles and fingers wiping away tears. As the school assembly applauds, so do the families at home.

Unlike our earlier examples, there is nothing particularly unusual or ambiguous about this story or these tears. The tale of a young man overcoming adversity is carefully constructed to generate (legitimate) feelings of joy and triumph and the audience at home responds to this pattern of feeling. But *Gogglebox* inserts a mirrored lens into this relay and sets up a series of reflections that emphasises points of commonality and minimises disruption. For the *Gogglebox* cast, recognition and empathy is found in a variety of reflections and identifications offered by *Educating Yorkshire*: a school girl is shocked to see Musharaf studying the same poem she read in class that day; a retired teacher responds to the value of the profession; a British Asian family celebrates the emotional openness of the young British Asian lads in the assembly hall. There is the sense in which the cast's own, often unspoken, histories are carefully revealed as underpinning their tearful reactions. However, closing down the range of responses to (just) tears opens up judgement on the manipulation of emotion as the programme makers explicitly direct the spectacular display of tears through an emotive musical score and multiple close-ups of the crying cast. This is less about the skill of the performers to convince as to their sincerity than to establish the frame of the performance and to allow for the articulation of specific cues. Crying, in this instance, is utilised to emphasise consensus rather than contestation across both the *Gogglebox* cast and by implication the audience at home, watching others watching television.

In a long and influential discussion of emotion in a legal context, Susan Bandes recognises emotion as an inevitable and even beneficial element of judgement.[8] But she also warns that we need to be aware that empathy – which we would suggest is a channelling or rationalising of our own emotional response to others' distress – is not neutral and it may often be conservative:

> In order to bridge disparate types of experience, so as to facilitate empathy across a broader range of contexts, it is often necessary to emphasize commonalities, and to downplay perspectives that are not shared. This may effectively serve to perpetuate, rather than challenge, the status quo (1996: 375).

The question that our example from *Gogglebox* asks is whether the consensus that emerges around tears dampens the differences between Musharaf, the cast members and the viewing audience, and in doing so obscures aspects of the programme that we might otherwise wish to challenge: its dependence on the ideology of the family (or kinship) and on a constructed hegemony of emotion which all but insists that there can be only one acceptable response to the tears we see on-screen. In this instance, in which real adversity appears to have been overcome and which speaks (cleverly, deliberately or coincidentally) to a range of different kinds of experience, this exhibition of contagious crying would seem to be benign. Yet if the same narrative and emotional manipulation is used for other purposes we may feel less comfortable. Equally, we might ask whether the shared crying is really enough: as hinted at by the cast's initial response and made evident in the narration, Musharaf was continually bullied and excluded by his fellow pupils (some of whom are presumably those we see crying at the assembly). Are their tears now really enough? The ambiguity of tears again makes us question whether the tears are 'real' or, perhaps, sufficient – is the emotional outpouring that we witness provoked by sympathy or shame, or is it simply the result of embarrassment and peer pressure? Wherever they spring from, the appearance of tears here usefully shuts down any further interrogation or exposure to the intolerance and bullying experienced by Musharaf, and while the story makes him (and his dedicated teacher) the 'heroes' of his story, in doing so it conveniently occludes the guilt of his peers for failing to tolerate and support his difference in the first place.[9]

Crying and tears as aspects of performance are, we have suggested, both mundane and marvellous. As material evidence, tears appear as visible cues for television audiences who then make quite complex and often contradictory interpretations of instances of crying that are increasingly commonplace across a host of television genres. While we have focused here on the 'performance' of tears in reality formats and argued that crying is perhaps one of the most significant yet murky skills of the reality show performer, the mystery of tears means that even when their meaning is continually negotiated and contested they retain the power to move us. By making an exhibition of themselves, reality show performers expose an ontological and epistemological instability that – we would argue – speaks directly to our current conception of self. In fact, it is through the productive incoherence of tears that we might further claim that television, in its ordinariness, continued sociality and communicative promiscuity, provides an unequalled platform to explore the messiness of contemporary subjectivity.

Notes

1   While animals can and do cry, the cause is predominantly understood to be physiological. Both Tom Lutz (1999) and Michael Trimble (2012) open their studies of crying by emphasising that the phenomenon, as an emotional response, is unique to humans.

2   For Robyn Warhol, in her study of crying, effeminacy and the narrative forms of popular culture, this division presents itself as a 'sincere and authentic emotional experience' versus the 'false sentimentalism and affectation' of mass, popular and feminised culture (2003: 11).

3   Wood, Skeggs and Thumin usefully link reality television's 'playing out of moral dramas' within traditions of earlier (feminine) forms of television, specifically the soap opera and the domestic sitcom (2009: 136–37).

4   Warhol writes that 'this theory of "catharsis" has been taken in the twentieth century to mean that audience members' bodies or psyches contain a given quantity of pity and of fear, and that the experience of weeping at a tragedy constitutes the "proper purgation", the healthy and controlled venting or draining of these emotions' (2003: 15–16).

5   We might also draw a specific link here with a desire for immediacy and the phenomenon of 'corpsing' (where the frame of performance is broken by the laughter of the performer). As Karen Lury has previously argued, 'corpsing engenders a moment where the television performer reveals his or herself as truly live, uncontrolled and expressive [...] it is this process of revealing that the audience almost greedily looks for, or hopes for, in much of television. For it suggests that form of direct communication, the existence of a real bond between performer and viewer, which television seems to promise, yet which it can rarely deliver' (1995: 127).

6   See Skeggs and Wood for discussion of *Schadenfreude* as an audience response to reality television (2012: 160–62).

7   As an art historian, Elkins is interested in trying to discover when, how and why people cry at paintings, based on over 400 letters he received. He recalls how, at first, the bewildering array of responses (and the general disdain he had for his topic from his fellow art historians) made him doubt his project.

8   Bandes' essay discusses the significance of the inclusion of emotive victim impact statements in capital murder cases. She argues that 'Emotion and cognition, to the extent that they are separable, act in concert to shape our perceptions and reactions. But more than that, much of the scholarship posits that it is not only impossible but also undesirable to factor emotion out of the reasoning process: by this account, emotion leads to a truer perception and, ultimately, to better (more accurate, more moral, more just) decisions' (1996: 368).

9   It is notable that the episode of *Gogglebox* falls back into intolerance in a moment of bathos that concludes the sequence: gay couple (later friends) from Brighton Chris and Stephen remark, 'It ain't often I have a little tear [...] mind you, you wouldn't want him to read you a bednight story would you?! It'd take all bloody night!'

8

Comedy, Performance and the Panel Show

Alex Clayton

The notion of the carnivalesque, most famously mobilised in Mikhail Bakhtin's discussion of the social institutions that gave rise to the work of François Rabelais, has long informed our sense of what comedy has absorbed from its historical precursors (Bakhtin 1984). It is almost a truism to invoke the idea that comedy, with its topsy-turvy logic and suspended morality, generates something like a festival atmosphere. But *how* this effect is concretely achieved, what we might call the craft of comedy, is less well appreciated. Indeed, much comedic performance, most especially group comedy premised on improvisation, takes pains to conceal its craft, sometimes even fabricating the impression that we are simply witnessing a bunch of friends messing around. This chapter is an attempt to develop a better grasp of that hidden work by identifying avenues of possibility and means of comic expression available to the seasoned improviser, specifically in the format of the TV panel show. It is also a demonstration that the nascent genre of close appreciative analysis in television studies need not restrict itself to varieties of fiction. Falling outside the categories of drama, documentary, and factual entertainment, the panel show – like the quiz show and the talk show – has drawn little academic attention (Mills 2015: 110) and has inspired no sustained passage of theory or criticism. That is what I intend to offer here.

Analysis of screen performance tends to rely on a distinction between invention and execution: typically, the crafting of the line of dialogue and the act of delivering it. The habitual focus on screen drama means performance tends to be equated with delivery. However, this is barely tenable when discussing improvisation, where invention *is* the execution, or at any rate they are not easily distinguishable (invention is, at least, not an event that has appreciably preceded execution). Moreover, the word 'performance' contains a shadow notion of putting on a front, dissembling or pretending in some way. For instance, it does not sit easily with such behaviour as spontaneously laughing or being surprised, because calling this a 'performance' might imply (even if we do not want, or if it is not important, to suggest) that it is somehow faked for entertainment purposes. (Making your laughter *available*, or *showing* yourself surprised, might be better ways of putting it.) These are two reasons why an analytical appreciation of improvisatory performance is challenging to undertake. What follows is an attempt to delineate some of the comedic activities undertaken by performers in this context. They might be thought of loosely as performative modes of comedy, although the distinction between them is more analytical than practical. In Robin Nelson's terms, I am offering not 'know-how' but 'know-what' (2007: 37). This is not a guide to being funny, nor a description of choices consciously entertained by performers in mid-flow. Seasoned comedians combine and move intuitively between these modes as professional cyclists shift unthinkingly between gears. It is nonetheless incumbent on the analyst of comedy to know and be able to say what its practitioners are doing.

Before that, a few words may be needed about the genre of the panel show. A direct descendant of the Victorian parlour game, it remains strikingly popular in a British context. The panel show – 'in which a number of celebrities (often comedians) answer questions or carry out tasks on a particular theme' (Mills 2015: 109) – was initially dominant in the UK on radio, as indicated by the success and longevity of such BBC Radio 4 shows as *Twenty Questions* (1947–76) and *I'm Sorry I Haven't A Clue* (1972–). Programmes such as the word-definition game *Call My Bluff* (BBC, 1965–88), the charades-based *Give Us A Clue* (ITV, 1979–92), and the topical news quiz *Have I Got News for You* (BBC, 1990–) were forerunners of the genre on television. Its potential as a comedic platform for free-wheeling improvisation, held in check by rules and restrictions, was perhaps most vividly demonstrated by *Just a Minute* (BBC Radio 4, 1967–) and *Whose Line is it Anyway?* (Channel 4, 1988–99). This handful of shows represents the most obvious inspiration for a genre that has mushroomed in British broadcasting post-2000.

The format I nominate as contemporary exemplar is *Would I Lie to You?* (BBC, 2007–), and the particular episode from which I will draw examples is the opening episode of the show's fifth season, first aired on 9 September 2011.[1] The format is simple: in front of a live studio audience, contestants arrayed in two teams (led respectively by comedians David Mitchell and Lee Mack) read out or recite a claim about themselves, typically something they have supposedly done in the past, and/or their relation to an object or person; the opposing team interrogates this claim by asking questions and querying the response; then, on the basis of the claimant's elaboration, the team must judge whether he or she is telling the truth. There are several rounds in which points are scored, and winners are commended at the end of the show. However, as is standard in panel shows, nobody cares about the scores. The sole purpose of the competition is to provide a platform for fun.

The main activity of the show is the elaboration and querying of licensed fibs and disguised truths. There are various pleasures available in this. Since the veracity of the claim is not revealed to us until the end of the round, there is the option to play along and try to discern the bluff (a deliberately weakly supported truth or an insistently supported lie) from the double-bluff (an insistently supported truth, or a lie so incredible in its details that, perhaps, 'it could only be true'), or otherwise to catch out the claimant in a contradiction or embellishment too far. Similarly, sharing ground with the talk show, there is the potential for mild pleasure to be found in celebrity self-exposé, at least mild suspense at its possibility, as when an embarrassing story is revealed as a true disclosure. However, the primary appeal of the format is the support it offers for improvised exchange and repartee.

As with any panel show, the choice of line-up is therefore an important factor. Comparable to the performance dynamic of a rock band or a football team, any particular line-up will yield different possibilities of collaboration and interaction, and potential for clash and overlap. Engineering this chemistry, so far as it is possible, is a matter of trying to find a sparky dynamic between different personalities, temperaments and voices. This is one reason, perhaps, for the use of regular 'captains' who provide some constancy in the selection and arrangement of teams. In this respect, season five of *Would I Lie to You?* builds from a solid base. Its two regular team captains, Lee Mack and David Mitchell, have by this point in the series developed a rapport, having settled into a dynamic (perhaps inspired by Ian Hislop and Paul Merton on *Have I Got News For You*) that emphasises their contrasting performance styles (for instance, Mack's quipping vs Mitchell's trademark ranting), persona type (extrovert vs introvert), vocal quality (raucous vs nasal) and social background (northern working class vs southern upper middle class). These contrasts lend texture to the interaction between the team captains as well as providing source material for light-hearted teasing and mimicking. On the whole, however, the show's brand of humour is rather gentle, more collaborative than confrontational, as reflected in the set design which places the teams in stalls at an obtuse angle to (rather than opposing) one another. In the centre, the host, Rob Brydon, is placed at a desk that joins the two stalls, forming a shallow arc. Brydon's persona anchors the show by complementing without overshadowing either of the two main performers. His dryness of delivery, for instance, strikes an affinity with Mitchell's sardonic humour, while his warmth and gregariousness chimes with Mack's natural confidence. The other guests are picked and positioned around this triangle of big personalities.

In the episode I've chosen to discuss, there is a definite PAC (Parent–Adult–Child) strategy to team composition. Mack is joined by sitcom star Miranda Hart and TV personality Nick Hewer (then an advisor on *The Apprentice* (BBC, 2005–)); this combination offers an enjoyable contrast between personae that are respectively vivacious (with a hint of the bubbly schoolgirl) and sober (shades of the stern headmaster). On the other side, Mitchell's team comprises the comic actor Rebecca Front and the comedian Jack Whitehall. Front's dry wit is underused here; for whatever reason, she is granted little screen time in the final edit and her main function seems to be to provide a kind of grounding presence, slightly mumsy against Whitehall's cheeky-boy act. Whitehall, on the other hand, is given quite a lot of screen time, despite being a risky choice for Mitchell's team, since his posh persona threatens to usurp David's usual position as the object of wisecracks aimed at his well-to-do background. (Indeed, at one point Brydon observes that Whitehall is 'making David look positively working class': 'I barely need to be here this week', Mitchell replies.) As a tactical counter to this potential overlap, Mitchell and Whitehall's performances emphasise differences in age and sexuality, through gestures and attitudes such as Mitchell's crossed-armed sensible questioning as against Whitehall's effusive hand movements and occasionally 'camped-up' poses.

I now move into trying to define some of the characteristic activities that comprise comedic performance in a group context, with reference to the focus episode – although I offer these activities as exemplifications relevant to other episodes, other panel shows, and, beyond that, to countless other modes of comedy and varieties of sociable occasion.

Magnifying

In service of the aforementioned group dynamic, one activity repeatedly undertaken to comedic effect in the episode is to amplify aspects of a persona through speech and gesture. For instance, when Whitehall casually refers to his mum as 'Mother' – an archaic appellation with aristocratic connotations in a contemporary British context – he closes his eyes in embarrassment upon seeing Mack's keen grin: the image is cut back and forth across the eyeline to convey this, and the studio audience greets it with laughter. By way of recovering from this slip, a mortifying betrayal of excessive poshness – or, more importantly, by way of turning a comic profit from it – Whitehall then plays up both elements, the poshness and the mortification. First he repeats 'He knew me through Mother!' in an accentuatedly patrician accent, then he puts his hand over his eyes, demonstratively leaning over and turning away, almost sinking below the level of the desk. This little piece of improvisation works to prolong the humour of the moment by stretching the distance between 'who he is' and how he 'wants' (or rather, doesn't want) to appear. It also contributes to the sociability of the occasion by acknowledging the values that prompted the laughter, rather than, for instance, going for a comically hardened defence that admits no shame (e.g. 'That's what I call her, OK?' – which might have worked on a more combative panel show). The business nudges his persona into the more endearing territory of 'posh-despite-himself' rather than stubbornly 'posh and proud' (or mock proud). The latter might have worked in a different context, but the former is more fitting for a show that is, after all, built around the prospect of the telling giveaway.

In the course of the episode, each performer (with the exception of Rebecca Front) finds the opportunity for self-magnifying a distinguishing trait. For instance, when business advisor Nick Hewer is given the task of persuading the opposing team that he and *Apprentice* boss Alan Sugar 'wind down' after filming by playing table tennis on the boardroom table, he uses the occasion to magnify the characteristic scrupulousness of his TV persona by noting, straight-faced, that the boardroom table is, in fact, 'slightly bigger than regulation size'. Hewer is not principally known as a comedic performer, but his talent for deadpan is apparent in the gesture with which he punctuates the line, a delicate scratch of an eyebrow, using ring finger and little finger to 'cover the lie' by suggesting false modesty in the boast. David Mitchell, meanwhile, aware that his persona incorporates the pedantic and may verge on the supercilious, plays an aside when interrogating Hewer about the logistics of the impromptu match. Having observed that 'you have to back off quite a way when playing table tennis properly', he seemingly checks himself mid-gesture and turns to the studio audience to add: '… I happen to know'. The aside foregrounds his propensity for smugness by acknowledging self-deprecatingly that he has just claimed common knowledge as specialist knowledge. It also magnifies another element of his persona, that of the unpractical bookworm, by implying that his sporting knowledge begins and ends with ping-pong.

Riffing

No small part of the comedian's craft in a format such as the panel show is to spot latent comic possibility in passing detail and work with others to help tease it out. Like a jazz group's extemporisation around a melodic phrase, comedians engage in collaborative 'riffing', the development of comic potential through features such as echo, analogy and extension. One such occurrence takes place during the round where Jack Whitehall is defending the unlikely (although, as it turns out, true) claim that he was once commissioned to paint a portrait of Gyles Brandreth's cat. Miranda Hart senses some potential for anthropomorphic suggestion by asking Whitehall if the cat 'posed' for him, to which Whitehall replies that he was required to do live sittings with the cat rather than work from a photograph. This generates some hyperbolic incredulity from Hart as she mocks the idea that a cat would be so compliant as to hold a photogenic pose for three hours, and to illustrate her point she does an impression of a posing cat. Hart renders her life model with front paws tucked under its chin such that it seems like it's *trying* to look cute – a touch of animal vanity – although in combination with the upright pose and glassy eyes there's also more than a hint of stuffed meerkat. This frozen quality of her impression prompts an exasperated counterclaim from Mitchell: 'You can totally do a sitting with a cat, cats are very sedentary!' he exclaims. 'They stay in one place all the time! You can't *command* the cat, but you say, "Oh here's a good moment, it seems to be…" [trails off] It's not a *wasp*!' Mitchell's objection picks up two threads from Hart, namely mobility and obedience, and tests them for further comic potential. The humour in the fleeting idea of 'commanding' a cat, which Mitchell accompanies with a regally sweeping hand gesture, has something to do with picturing the hubris of a painter trying to boss around his animal subject. Going in another direction, the equally brief juxtaposition of cat and *wasp* is pointillist improv of the first order, capturing not only an extreme of perpetual motion (in contrast to which the notion of a cat being 'active' seems absurd) but also pitching a cherished pet against a reviled pest of whom one wouldn't commission – and couldn't produce – a portrait.

Jousting

As we see from the above example, the adversarial set-up of the panel show gives the impetus for bouts of comedic mock-jousting: the attempt to outdo, undermine or knock down your opposite number, all in a spirit of festivity and, in a distant echo of the medieval court, with the blessing of the presiding host. One such bout takes place in the 'This is My …' round when Lee Mack is assigned to support the (false) claim that he once cut off the female mystery guest's ponytail on the school bus, having mistaken her for his friend Paul. Mack gets into trouble by trying to cover for the impression that the guest is considerably younger than he is, with the 'clarification' that they were in fact on the way to different schools. In true jousting spirit, Mitchell seizes on this weakness and tries his advantage, asking Mack to clarify that he was therefore on a public bus and had attacked a young girl he didn't know. Mack picks up on Mitchell's tone and remarks that it's starting to sound like a court case, telling him to 'back off'. Trying another tack, Mitchell picks at Mack's assertion that he had intended to cut off Paul's ponytail because he had found it annoying, adopting a highhanded tone to lecture him: 'it's up to people how they have their hair; it's not up to *you*, is it, Lee?' Mack immediately deflects: 'Trust me, David, if it was up to me, you wouldn't be having your hair like that.' Mitchell pauses a moment with crossed arms and a smile that declares no wound, then unfurls his hand as if returning a tribute: 'Likewise.' It may be a cliché and misstatement to say that comedy is 'all about timing', but the timing is essential here, both of Mack's near-instantaneous bite-back and the steady beat Mitchell takes before neutralising it. The pleasure is not just in quick wits but in a battle of wits, where that battle has an evolving shape governed by comedic intuition (and supported by predictive back-and-forth editing). Mock-prosecution gives way to mock-reprimand before getting mock-personal. Jovial badinage takes the guise of a slanging match, and, in a final twist, an insult takes the guise of a compliment.[2]

Conjuring

The shifts enacted here are bound up with another important aspect of comedic activity that might be called transformative suggestion. To take an instance from the focus episode, Lee Mack has challenged Jack Whitehall to prove his artistic credentials with pen and paper in support of the cat portrait claim. The disruption of the normal terms of engagement, as Whitehall takes quietly to the drawing task beneath his desk, is marked by a direct-to-camera statement from the show's host, Rob Brydon: 'I should at this point, uh, tell viewers at home that, uh, whilst we do like to receive your paintings, we can't return any of them.' While a somewhat obscure connection, for viewers of a certain age who grew up watching children's television in the UK the line will momentarily transport them back to a childhood sense of bafflement at this apparently churlish broadcasters' rule. As team captain David Mitchell, sitting beside Whitehall, observes his teammate's progress beneath the desk, Lee Mack pipes up and we cut over to his team for the duration of his remark: 'People who have just turned over are gonna be thinking "What on earth is David looking at?"' And indeed, on return to Mitchell's team we see an image transformed – and Mitchell compliantly plays along, cocking his head and folding his hands, gazing at Whitehall's lap with an expression that is two-parts innocent to one-part sexually curious. Mack's interjection conjures a visual gag out of dead air (Figure 8.1).

Further instructive instances of transformative suggestion can be found when David Mitchell is charged to defend the (false) claim, 'I killed a rat with my BAFTA'.[3] Mitchell renders a scenario wherein he spots a rodent loitering by some bin bags and instinctively grabs the coveted statuette from a bookshelf. Mack spies an opening (see Jousting, above) and asks for confirmation, with a wolfish lick of the lips, that Mitchell keeps his BAFTA 'on display'. Rather than deny the implication of vanity, Mitchell shifts mode to

Figure 8.1 David Mitchell (right) keenly observes whatever it is that Jack Whitehall (left) is doing under the table (*Would I Lie to You?*, BBC, 2007–)

sarcastic amplification (see Magnifying, above): 'Oh absolutely, it's got lights round it…' He trails off but the suggestion is given momentary volume by some impressionistic hand movements that sketch the impressive size of the non-existent display case. While the ostensible target of the joke is Mitchell's vanity, his comedic intuition knows the humour to derive from maximising the distance to be traversed by the imagination, from picturing the BAFTA as a lumpen object so unloved that it can serve to kill a rat to picturing it as a cherished artefact and basis of an elaborate shrine. Having made them incompatible – if the BAFTA is in a cabinet in Mitchell's 'me-room', it is hardly ready-to-hand – these notions of the sanctified and the disposable are then incongruously fused together, mere moments later, with another improvised gesture. In a visual demonstration of his eccentric rat-killing technique, Mitchell describes and shows how he held the award at arm's length, hovering it above the rodent – rather than, say, hurling it. The gag remains fleeting and unembellished, but I gather the logic was, picking up on the idea of it as a holy object, to evoke the notion of the BAFTA as crucifix. The conceptual density of this suggestion, with its melding of the glamorous and the occult, the modern and the medieval, where hunting a rat is akin to dispelling a demon (unseen, unwanted, beastly), together with its sheer impracticality as a method of pest control, gets an appreciative response from Mack. I offer it merely as one element of an overall comic texture where hundreds of ideas and images are conjured, modified, pushed apart and brought together – so densely woven as to go virtually unnoticed.

Playing Up

In a panel show, the conventions of the format provide the basis for extemporisation and repartee, the ground for flights of fancy. For instance, the taken-for-granted structure of having a succession of different rounds (marked by formal introduction and wrap-up via cards and/or teleprompter) is important for giving the show an underlying rhythm and for manufacturing a periodic tone of 'settling down'. Without the end-of-round prompt and cut-off, escalations of comedic energy could only end with the inherent disappointment of fizzling out. Furthermore, rounds ensure concentration, both in the sense of providing a literal statement on which minds are focused and in the sense that the discussion will end up distilled as a mere few minutes of screen time, enacting a healthy pressure on contestants to make their ruminations and observations snappy, compression at any rate being generally favourable to humour. Finally, the conventions of the game temporarily suspend and in part supplant ordinary conventions of social nicety, such that contestants are effectively licensed, for instance, to insult and shout at one another. This is important for panel shows not because rudeness with impunity is inherently funny (in a sense it is not rude if it is licensed), but because it fosters a mode of comedic exchange less encumbered by the fear of hurt feelings and the delay of thinking twice.

But there is more still to the panel show's insistence on structure and to its constant reminders of its routine ways of proceeding. The pleasures of screwy digression or madcap fixation on irrelevant detail, for instance, can only emerge against a baseline sense of how things are normally, logically, or most directly done. As a simple example, during the cat portrait segment in the focus episode, Nick Hewer reports that he happens to know the wife of Gyles Brandreth (the supposed commissioner of the portrait) and asks Jack Whitehall to confirm her name. With a sideways look, Whitehall names her as Michelle, which could either be a factual report, a wild stab in the dark, something that by chance Whitehall happens to know even though the story is false, or (if he were canny and wanted to throw them off the scent) a deliberately incorrect name even though the overall story is true. In response Hewer gives a tight smile and with a dismissive shake of the head turns to his team captain, Lee Mack, who looks back at Hewer in expectation and asks: 'Do *you* know Giles Brandreth's wife?' Hewer shakes his head no. The humour in performance comes from the constancy of Hewer's expression, as the gesture of an apparent breakthrough imperceptibly gives way to back-to-square-one resignation with the realisation that there was, as it turned out, no way to verify Whitehall's answer. But there is also a satisfaction in the inventiveness itself, in recognising the initiative to turn the tables on who is expected to bluff, on the game's convention of earnest enquiry. That the parameters of a game are after all malleable, that there are countless ways of playing it, is one of comedy's perpetual affirmations.

For this reason the 'rules' of a panel show are necessarily baggy, open to question, and much comic mileage can be made from the querying and bending of what can seem quite arbitrary and fluctuating rules.[4] Despite this, or alongside it, there is the counter-requirement for there to be some figure of authority or maintenance of order against which contestants have the opportunity to be unruly. This figure typically takes the form of the host (for instance, the schoolmasterish Stephen Fry in the long-running BBC series *QI* [2003–]), but in *Would I Lie to You?*, where the host Rob Brydon is often quite silly and has his own turn at being a claimant, the baton of authority is, as it were, passed around between the various contributors. Part of the task of effective performance on this show is therefore to recognise when it is time to wield the authority stick. David Mitchell, for instance, has a good line in stop-this-madness bluster that places him in turn as a potential comic butt. Despite his trademark cheek, Lee Mack can also be found, on occasion, trying to bring things back on track. In the focus episode, it is very often Nick Hewer who adopts the role of 'the grown-up in the room' – for instance, when being quizzed about the aforementioned table-tennis-in-the-boardroom claim. When Rebecca Front asks Hewer where he and Lord Sugar kept 'the bats', Lee Mack gets in quickly with 'She's left now, hasn't she?', a sly reference to the formidable Margaret Mountford, Hewer's former co-advisor on *The Apprentice*. As the studio audience responds with laughter, Hewer turns to ask him sternly if he is referring to Margaret. (There is some well-judged editing at this moment that presents his face in profile, accentuating the accusing turn.) 'Not Margaret, no no, no no, not her,' Mack replies, disingenuously, shaking his head, as if he's been caught out by a schoolteacher. Then, when Hewer turns back to engage the question, Mack – encouraged by the audience response – shifts immediately to nodding and mouthing to the opposing team something like 'Yes, Margaret, I'm talking about Margaret ...'. It's as if Mack has become a naughty schoolboy once again, and makes us complicit in his mischief (Figure 8.2).

In fact, this little business engages all the modes or aspects of comedic performance I have so far discussed in this chapter. Mack's behaviour here *magnifies* both the impudence of his persona and the sternness of Hewer's. It *riffs* on the audacity of the original interjection by play-acting duplicity and by the gratuitous confirmation of what we well knew to be Mack's intention. It *jousts* with Hewer by undermining his claim to authority, and *conjures* the scenario of a detention schoolroom where a disruptive pupil is mouthing behind the teacher's back. Finally, it *plays up* against the mock-authority of Hewer and gently teases the convention of the game whereby team captains are expected to side with their own side and conspire against the opposition.

Figure 8.2   Lee Mack's impish grin (*Would I Lie to You?*, BBC, 2007–)

The Image of Community

In addition to all of this, Mack's naughty schoolboy turn exemplifies a characteristic movement, in panel show performance, and in ensemble comedy more generally, between what we might call centripetal and centrifugal playing. Any seasoned troupe performer knows that a balance must be struck between 'playing out' (performer to audience) and 'playing in' (performer to performer). There is arguably no pure form of 'playing in', of course, so long as the engagement between performers is staged and offered for the audience's view. However, there is a risk that extended internal playing, especially when it features in-jokes or indulges troupe interests, almost to the exclusion of the audience, can be read as a kind of 'turning away'. Conversely, too much 'playing out', exemplified by direct or near-direct address, can make a panel show seem too much like a series of individual variety 'spots' or miniature stand-up routines. This runs the risk of too heavily reinforcing the distinction between the virtuosic performer and the audience 'out there'. This may be appropriate for some kinds of comedy, but it runs counter to the ethos of this particular variety. It is, for instance, important to the panel show format that its participants are shown to be having fun, becoming a kind of internal audience to whatever is unfolding around them. The proliferation of reaction shots constructed through the editing in *Would I Lie to You?* offers the pleasure of seeing one comedian, momentarily placed as connoisseur, enjoying the work of another contestant as she or he builds on what has come before. The vision of a mutually appreciative community is the format's most direct appeal to utopian feeling (Dyer 1979). This, along with the abundance of teasing and good-spirited fault-picking, projects the image of something like a family. The alternation of 'playing in' and 'playing out' is what issues the invitation for us to join that community, to complete the circle of participants.

The forging of that invitation is exemplified, in the focus episode, by some extended shenanigans featuring an item of oversize clothing. It is typical of the show's egalitarian spirit that its host, Rob Brydon, has a turn at making a truth claim. On this occasion he brandishes a huge bright orange sweatshirt, claiming (with slightly robotic intonation, as if feigning embarrassment, at any rate offering a tonal counterpoint to the cuteness of the idea) that he and his wife wear this garment together when 'cosying up on a chilly evening'. They call it their 'cuddle jumper'. There is studio laughter here and a cutaway shot to a laughing Jack Whitehall, briefly gazing out in delight and disbelief to the live audience, screen-left, before channelling our attention back to Brydon. This cutaway is a small detail, an intuitive editing selection, but important for the utopian feeling mentioned above. Whitehall here appears to share an exchange with the audience, laughing with them, as one of them, one of us. His out-of-frame exchange invites us, via a kind of double surrogacy (the performer for the studio audience, the studio audience for the television audience), to consider ourselves passive participants in the moment, much as he is, and not as dislocated viewers.

The editing now alternates between reaction shots of all the contestants – leaving no one out, in a gesture of inclusivity – and shots of Brydon as he unfolds the sweatshirt and displays it frontally (for the home audience via the camera, and for the live audience behind the camera, a choice that further unites their perspectives). His unfolding finally reveals that the cuddle jumper has two holes for two heads. Among the various reaction shots that accompanies the ensuing laughter, we might notice Miranda Hart turn instinctively to the audience as she giggles and claps along with them. Lee Mack asks for a demonstration of this artefact, and volunteers his teammate Nick Hewer (who, true to his persona, feigns reticence) to model the jumper with Brydon. The humour in the sight of this pair sporting the bright orange sweatshirt – Hewer perched on Brydon's lap, as Hart observes, like a ventriloquist's dummy – largely comes from Hewer's performance of compromised dignity. But the highlight is the set of reaction shots of David Mitchell, howling with laughter at this spectacle, unable to contain himself. The festive atmosphere is capped when, in a moment of ostensible improvisation, Lee Mack cuddles up to his teammate Miranda Hart inside *her* sweater and they miraculously start speaking at the same time, like cartoon conjoined twins who complete each other's sentences (Figure 8.3).

If there is an element of serendipity about this final surprise, the path has been prepared by virtue of many astute choices, from casting, to team distribution, to editing, to the choice of stimulus material for the round. Above all, however, the achievement of the panel show, when it works, comes from the trust placed in its performers, and the way their comic instincts are productively channelled and combined to form the image of a creative, teasing, laughing community. I hope in this chapter to have made a start towards a suitable appreciation of the activities that comprise this craft, which has for so long gone unrecognised: the *construction* of carnival, the art of merrymaking.

Figure 8.3 Rob Brydon (left) plays Tweedledum, Nick Hewer (right) plays Tweedledee (*Would I Lie to You?*, BBC, 2007–)

Acknowledgement

I am grateful to the editors of this volume for their attentive comments on an earlier draft of this chapter.

Notes

1   Limiting discussion to a single episode allows my observations to be more readily check-able, restricts cherry-picking, and illustrates the dexterity with which the perform-ers in that episode move between different comedic modes. As at the time of writing the episode is available to view online: www.youtube.com/watch?v=paKvWxTNl_w (accessed 26 April 2019).
2   The notion of 'incongruous intentions' as a manifestation of the comic may have perti-nence here. (See Clayton 2012: 52–54)
3   BAFTA stands for British Academy of Film and Television Arts, shorthand for an award bestowed by that organisation, although the phrase 'plays' much better using the acronym as shorthand.
4   This is not so evident in *Would I Lie to You?*, perhaps, but one need only think of Radio 4's *Just a Minute*, and of the weekly rule questioning and discovery of loopholes that has been a constant of the show for more than 50 years.

Part 3

The Television Performer

9

An Actor Diversifies: A Diachronic Examination of the Work and Career of Tony Curran

Gary Cassidy and Simone Knox

This chapter examines the work and working practices of Tony Curran, an actor we consider to have noteworthy versatility. Focusing on Curran's career in television across a number of projects, contexts and years, our argument is that his considerable versatility involves a number of factors, both within and exceeding his creative agency. These include confidence and creative risk-taking; an emphasis on acting as a craft that requires discipline, preparation and attention to the methods of production and exhibition; and awareness of the social culture within which the careers of actors are located.

The chapter combines close analysis of selected moments of Curran's acting with insights gleaned from a personal interview with him, conducted on 6 May 2016. Brett Mills cautions that 'it is important to acknowledge the subjectivity within any interview material, and not to use this data as evidence of certain kinds of working practices' (2008: 152). We agree with his first point; but, especially for scholarship on acting, interviews can provide pertinent knowledge that would otherwise remain inaccessible, and we map Curran's thoughts against our study of the relevant industrial/production contexts. Our methodology further brings together scholar and practitioner perspectives, which adds to our analysis a valuable awareness of and emphasis on process and the reality of practice. This complements in turn the actor at the focus of our attention: with audiences more likely to recognise him by his face than by his name, Curran is the kind of actor who would not have typically been written about in past scholarship on performance which has traditionally placed an emphasis on stars. He is not a global A-lister, and thus still embedded within the precarious realities of the creative industries, but a professional actor at a level of success where he has played diverse roles in numerous projects across different national contexts without getting trapped in typecasting. This makes him an important and exemplary case study, with his achievements worth highlighting and illuminating for the study of acting.

Our analysis of his performances in *This Life* (BBC, 1996–97, 2006), *Doctor Who* (BBC, 1963–89, 2005–), *Defiance* (Syfy, 2013–15) and *Roots* (History Channel, 2016) will pay particular attention to how medium and/or genre may intersect with Curran's acting, and the preparation and rehearsal methods he draws upon to manage such diverse parts. This critical evaluation of one actor's working practices will pay due attention to the production and industrial/institutional contexts within which Curran's career has unfolded. Our analysis will therefore illuminate issues concerning acting and the professional life of actors within the contemporary creative industries that have wider relevance beyond the chapter's focus on Curran himself.

To introduce him in a little more detail, the career of Tony Curran is both representative of and somewhat atypical for contemporary British actors. Like many of his professional peers, Curran, who was born in Glasgow in 1969, trained at a drama school: the Royal Conservatoire of Scotland (the Royal Scottish Academy of Music and Drama at the time he received his Diploma in Dramatic Art in 1993). Since signing with agency Scott Marshall Partners following his graduation showcase, he has worked as a professional actor across film, theatre, television and video games. With his profile developing through industry recognition and acclaim, Curran is part of the wave of British and Irish actors who relocated during the 2000s to the United States, where he has played roles in several high-profile television series.

Given the contexts of training and professional experience, Curran's and his peers' work are located within a Western, broadly Stanislavskian approach to performance. With 'An Actor Diversifies' recalling the seminal *An Actor Prepares* (Stanislavsky 1964), this chapter's title indicates the centrality of Stanislavski in our approach to the analysis of Curran's acting. While there is no space here to discuss his complex and complicated work and legacy, we will note that, broadly speaking, Stanislavskian approaches to acting focus on realism, naturalism and 'emotional truth': notoriously unstable, slippery terms.[1] In the expectation that our readers are more likely to be familiar with the terms 'realism' and 'naturalism', we note that we use the term 'emotional truth' to refer to acting that imaginatively utilises an actor's personal experiences in an effort to produce a performance that is read as believable by an audience. Curran himself is one of many actors who see finding the truth of any given scene and making the performance truthful as their main intention and responsibility.

These commonalities with a number of his peers notwithstanding, Curran's professional development is unusual for two reasons. Firstly, there is the level of success and near-continual employment he has achieved since leaving drama school. This success is elusive for the majority of professional actors, and made more remarkable within the context of the current high-profile attention given to the working conditions for actors, which have been highly precarious in Britain for some time. Rosalind Gill and Andy Pratt have noted that creative workers experience 'short-term, insecure, poorly paid, precarious work in conditions of structural uncertainty' (2008: 2), and actors (in Britain and elsewhere) are no exception. Recently, the actor's union Equity estimated that 'around 90% of trained and qualified actors are out of work at any one time' (The Stage Castings 2015). Equity has furthermore articulated its concerns around low pay, as 'almost half of respondents in our most recent survey earn less than £5,000 per year from their professional work' (Equity 2015). In light of these figures, it is not surprising that Sam Friedman, Dave O'Brien and Daniel Laurison have likened being an actor, especially for anyone from a non-privileged background, to 'skydiving without a parachute' (2016: 10). From a poor working-class family, Curran's very decision to become an actor is perhaps the most compelling proof of the commitment to risk-taking that we identify across his career.

Secondly, Curran's career is unusual because the roles he has been cast in have been noticeably diverse. On television, since his first breakthrough as gay Scottish plumber Lenny in *This Life*, his roles have included villainous Orlick in the literary adaptation *Great Expectations* (BBC, 1999); an SAS soldier in ITV1 action series *Ultimate Force* (2002–08); a Russian torturer in US action serial *24* (Fox, 2001–10); Dutch painter Vincent van Gogh for British science fiction series *Doctor Who*; a medieval king in TV mini-series *Pillars of the Earth* (Starz, 2010); British politician Robin Cook in comedy-noir *The Hunt for Tony Blair* (Channel 4, 2011); an Irish immigrant labourer in 'quality' drama *Boardwalk Empire* (HBO, 2010–14); second in command of a biker gang in gritty drama *Sons of Anarchy* (FX, 2008–14); an alien in science fiction series *Defiance*; and a sadistic overseer at a slave plantation in the recent remake of *Roots*. This range is remarkable, especially considering the reality of typecasting that affects the working lives of many actors. Trevor Rawlins notes that it is

> far more usual in television production to cast actors close to their age and physical type out of economic necessity. [...] The logical conclusion of this industrial reality is that actors will increasingly tend to play within their own age and type range.
>
> (2012: 212)

Given that certain forms of typecasting may be in effect at different levels of the casting process (including producers, casting directors and talent agents, both individually and accumulatively), that Curran has played so many different roles reflects a very positive working relationship with Scott Marshall Partners, which continues to this day and is clearly a crucial factor in sustaining his career over more than two decades. It also suggests a wider recognition within the creative industries of Curran's aptitude for variety.

This Life: Creative Risk-taking and 'Invisible Acting'

To begin with our first case study, *This Life*, a drama series about a group of flat-sharing twenty-somethings in London, marked an important turning point in Curran's early career. After signing with his agent and making the (to some extent, inevitable) move to London, the roles Curran secured in the mid-1990s reflect many British peers' curricula vitae, namely one-off roles in long-running television genre programming such as *Taggart* (ITV, 1983–2010) – almost obligatory for emerging Scottish actors – and *The Bill* (ITV, 1984–2010). Having amassed a number of roles that were relatively non-descript, as reflected by the character names given in the credits (e.g. 'Travel Agent' for *Shallow Grave* (dir. Danny Boyle, 1994), 'Police Officer #2' in *Grange Hill* (BBC, 1978–2008)), Curran was cast as Lenny in the second season of the high-profile drama *This Life*, which would prove to be his first breakthrough. This part was actually written for him by scriptwriter Eirene Huston, an unusual occurrence for an actor at this stage of their career. It happened because Huston was a personal friend of Curran's;

however, she experienced resistance from the producers and Curran underwent a rigorous auditioning process. Curran has a high 'strike rate' with auditions, not only because he prepares thoroughly, but also because he is adept at sight-reading, having enjoyed improvising since drama school. Thus his casting here reflects ability, historical contingency and the significance of the social culture within which actors' careers develop – or do not develop. Much of what Tom Kemper has argued in his work on the rise of Hollywood agents from the late 1920s to the 1940s evidently still applies to the present-day creative industries (including in the UK), especially his following point:

> Crucial to my argument here is my conception of Hollywood as a business world embedded within a *social network* (and vice versa). This may not be big news, but it adds an important perspective to understanding the business, which [...] cannot be extracted from the *social culture* in which it is rooted.
>
> (2010: ix–x; emphases added)

Curran has long been aware of the significance of the social culture – he calls it a 'big circus' – within which acting is located, and he takes a very disciplined approach to this, especially as far as his emotional labour is concerned. Emotional labour here does not refer to the emotions with which actors engage for the roles they play, but to the work they undertake in terms of managing their own feelings during the processes of auditioning, rehearsing and acting, such as the pressure of feeling like an outsider when joining a long-running show. Elly Konjin points out that '[e]motions on the level of the actor-craftsman are rarely considered in traditional accounts of acting, except incidentally in relation to stage fright' (1995: 133). Nevertheless, as David Hesmondhalgh and Sarah Baker insist in relation to creative work in the television industry, such emotions 'cannot be detached from an understanding of the specificity of cultural production' (2008: 103). They form a significant part of the professional working lives of actors such as Curran, whose disciplined approach to his emotional labour includes him striving to cultivate positive working relationships without taking notes and feedback personally. This benefits his work on two levels: firstly, as Eric Hetzler has found in his study of actors and emotions in performance, if actors do not manage their emotions, especially any negative feelings, these may hinder them from connecting 'better to the task and thereby be[ing] "in the moment"' (2007: 75). By managing his emotional labour effectively, Curran avoids such a detrimental impact on his acting. And secondly, as he avoids creating negative, unproductive atmospheres on-set, he aids his future career prospects. As Hesmondhalgh and Baker further note, emotional labour is 'involved in developing working relations with team members that lead to those very important enduring contacts – contacts that lead to further contracts' (2008: 114). Aware of the difficult working conditions for – and the firm limits on the agency of – actors, Curran is one of many (if not all) members of the acting profession who know that the perceived mismanagement of emotional labour could damage his career prospects within the precarious social culture of the creative industries.

A scene from *This Life* episode 'From Here to Maternity' (season two, episode 15) demonstrates two of the reasons for Curran's breakthrough and why he is such a versatile actor: his willingness and ability to take creative risks, and the attention he pays to the methods of production and exhibition. In the scene, which consists of a number of different shots, Curran's Lenny, who has only recently arrived within the story world, is having a post-coital conversation with Ramon Tikaram's Ferdy about the fatal accident of a friend of Ferdy's. Curran's risk-taking is apparent in the fact that he chooses to play a gay character here completely naked (see Figure 9.1). While this is not entirely evident on the screen, as within the wider framed shots Curran lies on the bed diagonally across the frame with his leg pulled up, he decided to forego the modesty pouch for the purposes of realism and intimacy. That he chose to play the scene naked might at first not sound that noteworthy, but consider that Curran, a cisgender and heterosexual man himself, was only a few years out of drama school by this point, and still a relative newcomer to the *This Life* set (his character had only appeared in two earlier episodes). Entering and being new on a set is an anxiety-inducing experience for many (if not all) emerging actors, and this anxiety, Curran points out, still revisits him to this day (as it does for many of his peers). Trevor Rawlins (2012: 144) has noted that temporarily joining a long-running television show where the crew and regular cast are familiar with each other can present a number of challenges, especially the pressure of feeling like an outsider.[2] These are precisely the sort of parts which emerging actors tend to secure, and in the case of *This Life* a comparatively inexperienced Curran found himself on the set of the second season of the, at the time perhaps, most high-profile British drama series. Thus Curran's decision to be entirely naked is a striking

Figure 9.1   Ramon Tikaram's Ferdy and Tony Curran's Lenny connect in an intimate moment in *This Life* (BBC, 1996–97, 2006)

example of his commitment to risk-taking, demonstrating that his focus in the scene lies very much with creating and supporting a sense of intimacy, which has particular significance for the domestic medium of television in general (see Newcomb 1974: 245–48), and which the audiovisual style of *This Life* sought to evoke (see Cooke 2015: 191–95).

This sense of intimacy is further facilitated by the fact that Curran underplays the scene. By 'underplaying', we mean that he delivers his lines softly, and his physical actions and facial expressions remain restrained. Called for by the scene as scripted, this 'invisible acting' (within practitioner discourses on acting also known as 'small acting') while naked and playing against one's sexual orientation requires considerable confidence from actors, as it can feel like one is not actually doing much at all. The actor's inner experience here links to one of the paradoxes concerning the evaluation of acting in a contemporary Western context – where 'invisible' acting is frequently judged superior to 'visible' acting – summed up by Brenda Austin-Smith as follows: 'If you can see it, it isn't working; if it works, you can't really see it.' (2012: 20) To 'do nothing extremely well', which Alfred Hitchcock (in Naremore

1988: 34) famously cited as one of the hallmarks of an accomplished screen actor, requires perhaps especial confidence from emerging actors, who are often understandably keen to make the most of a given opportunity, and to demonstrate that they can act and 'do *something*'. It is significant to our argument that Curran has maintained a lot of confidence since his early years, having built his belief in his abilities at drama school, which was a 'very helpful' experience for him. The decision to underplay is not facilitated by his being naked in the scene, nor vice versa; both choices speak to the confidence supported by his training, as well as to his focus on emotional truth, which is present here in a commitment to developing a relationship between two lovers. Curran's confidence to 'do little' shows not only in his restrained use of body, face and voice, but also in the fact that he takes time to respond. For example, when Tikaram's Ferdy tells him, 'All it takes is for some arsehole to forget to check his mirror and you're dog meat', Curran lets the line land and considers it for a moment before replying with: 'Did you see it?' Supporting the sense of intimacy marking the scene, Curran's listening and lack of self-consciousness here[3] help to efficiently establish his character as insightful and empathetic, as well as demonstrate his commitment to creative risk-taking.

In this approach to the scene, his risk-taking is not only informed by his confidence, but also both aided by and responding to the particular production methods employed for *This Life*. These, as Lez Cooke (2015) has discussed, were shaped by the economic and ideological concerns of Tony Garnett, whose World Productions shot the programme on digital Betacam. As Garnett has noted:

> So we create a style which allows us to be very flexible in what we shoot [...] and we'll do everything we can to facilitate an actor being 'in the moment', in character. So actors are not put on marks and it's the camera's job to find them and to catch them. So that means, with that amount of time, there's no tracks, there's no dollies, it's steady handheld because then you can just adjust for the movement of the actors.
>
> (in Cooke 2015: 193)

Curran's experience concurs with this summary, and one of his preoccupations as an actor is to pay attention to the methods of production and exhibition, down to the level of detail as to what type of lens is being used for shooting. The scene under scrutiny is somewhat atypical for *This Life* in that it is quite static, with both characters lying on a bed; and the shifts between wider and tighter framings are realised through cutting instead of camera movement. However, it resonates with the more dominant filming methods for *This Life* in that the actors here do not have to be concerned about marks, camera framing and microphones. As a result, Curran was aided in his risk-taking by his knowledge that the production methods were privileging the actors and his 'invisible' (or 'small') acting, and not vice versa.

Doctor Who: Unconventional Preparation and Supporting the Production Methods

With his profile receiving an important boost from the success of *This Life*, Curran's career developed to include being part of the main cast for *Ultimate Force* (ITV, 2002–08) from 2002 to 2003, as well as being cast in parts of different sizes in US high-budget films (including a bigger role in *The 13th Warrior* (dir. John McTiernan, 1999), a minute part in *Pearl Harbor* (dir. Michael Bay, 2001) and a substantial role in *Underworld: Evolution* (dir. Len Wiseman, 2006)). This reflects the career currents of many professional actors, whose employment paths rarely follow a linear trajectory. Another milestone for him was the Scottish BAFTA and British Independent Film award that he received for his performance in independent film *Red Road* (dir. Andrea Arnold, 2006). This high-profile industry acknowledgement aided his relocation to the USA during the 2000s, when he started getting cast for a (by now) substantial list of high-profile and/or acclaimed US television series.[4] Since 9/11, the United States Citizenship and Immigration Services has required foreign actors who apply for a work permit to provide a file of evidence for a high level of accomplishment in the creative industries,[5] and Curran's prestigious awards for *Red Road* have certainly helped his transatlantic career. Like many of his fellow actors from Britain and Ireland who have established careers in the USA, he has combined this with selected projects in the UK. One of these has been his guest role as van Gogh in the *Doctor Who* episode 'Vincent and the Doctor' (season five, episode ten, first broadcast in 2010), in which the Doctor (Matt Smith) and his companion Amy Pond (Karen Gillan) travel in time to visit the Dutch painter and fight an invisible monster with him. As van Gogh, one of his few roles so far based on a non-fictitious person, Curran demonstrates his versatility through delivering a very expressive, emotional performance in marked contrast with his restrained work in *This Life*, drawing on unconventional preparation and working in support of the quite particular demands of the production here.

Very aware that this role needed a sensitive approach, especially as *Doctor Who* is a show watched by families, Curran began preparing for the part through more conventional means: he researched van Gogh's life and in particular his struggles with mental health, reading van Gogh's letters to his brother Theo and visiting museums in Austria (where Curran was shooting at the time) to see some of his paintings. Research and preparation form the backbone of Curran's approach to acting:

> There's always research to be done, and I research as much as possible. […] Then, when the cameras start rolling, you need to have looked into all that backstory for any character, and then you just have to start playing what you've rehearsed. I always say: 'if you fail to plan, you plan to fail', so I prepare as much as possible for a character.

Once production in Croatia had begun, he made use of a more unusual preparation method for the scene when the Doctor finds a distraught van Gogh in bed, unable to cope with the thought of having to say goodbye to his new friends. As Curran recalls:

> There was actually a night when I slept on the set and I drank some absinthe and went and slept on the bed in Vincent's room where we actually shot one of the scenes later, when the Doctor comes in and asks him if he's OK, and I come round and I scream at him and he's all emotional and upset.

Spending the night before shooting the scene on-set in the bed, Curran used affective memory techniques[6] that involved drawing on his own fears and painful memories concerning his family to help him access 'that place', namely the emotional and physical state required for the scene. This preparation shows itself in the shots that begin the scene, in that Curran is genuinely crying, his eyes puffy and his tears dripping off his nose. This is an emotionally truthful performance of van Gogh's distress, and with Curran's night prior to shooting bearing some resemblance to his character's night of painful thoughts, Curran's performance successfully addresses one note that directors often give to actors: 'Where has your character just been?'

For his preparation for the remaining shots in the scene, in which van Gogh rejects the Doctor's efforts to cheer him up, Curran's attention shifted to the needs of the production methods, which are quite particular, in that the set of van Gogh's bedroom is modelled after the painter's famous 'Bedroom in Arles' paintings, and the camera frames Curran on the bed in a long shot in order to place the character in his distress within the iconic image. The practical consequence of this for Curran was that, with his customary awareness of the methods of production and exhibition, he needed to ensure his performance in these shots would be 'big' enough to be sufficiently visible within this particular framing and already very visually stimulating backdrop. This meant that he had to use his whole body for any gestures, which is difficult when lying on a bed (especially a narrow one, as is the case here), as one's own body weight and mass work against this. Having plotted out the use of his body during his preparation, Curran uses his core muscles to twist and turn on the bed (see Figure 9.2), while stretching out his arms in a flat semi-circle, a gesture somewhat reminiscent of Greek tragedy performances, to convey van Gogh's despair. There is a certain theatrical quality to his movement, but this reads as believable as it helps to convey the force of van Gogh's rejection of the Doctor's

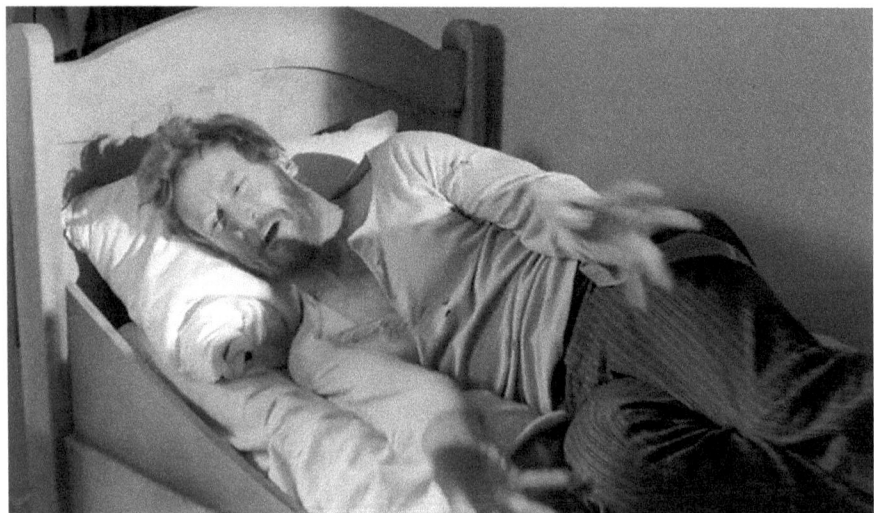

Figure 9.2 Tony Curran's van Gogh experiences a moment of total despair in *Doctor Who* (BBC, 1963–89, 2005–)

insistence on hope as well as the painter's mental distress, further linking to the common perception of van Gogh as a tortured genius struggling with intense feelings. In a very different working context to *This Life*, in which the production methods privileged the work of the actors, Curran's acting adapts successfully to support the particular needs of the scene and production methods.

Defiance: Diachronic Character Development and Animalistic Performance

Curran's acting in *Doctor Who* received positive feedback, not only from critics but also from viewers: a number of fans have subsequently approached Curran at conventions to express their appreciation of his sensitive portrayal of a character with mental health issues, which is why Curran considers this his most rewarding role to date. Curran continued to focus on work in the USA, signing with US agency Domain Talent and manager Tammy Rosen, and after several call-backs was cast in his longest-running role to date: *Defiance*'s Datak Tarr, a scheming Castithan alien who ruthlessly works to obtain legitimacy, status and power for his family. This role presented a number of acting challenges arising from the genre of telefantasy, such as frequently speaking some lines in the alien language that was created for the show and held no resonance for the actors, or balancing the more heightened acting style the programme employs with wishing to make the character believable. Given the pressures on schedules and rarity of formal rehearsals, it is now opportune to consider one of the determinants of television acting as identified by Roberta Pearson (2010: 167–68) and confirmed by Richard Hewett (2015: 74–77): time. We are specifically interested to examine how Curran responded to the opportunity to develop a character over an extended period, which television especially may afford.[7] As Curran notes:

> For me, the journey of a character in long-running television can be really interesting for an actor, because of the many different situations and many different experiences they have in their life; you can map the arc of that individual. And once the boxset comes out, you can go back over it, and you can see the arc of that individual. So, for an actor to play that is quite satisfying.

A good example of how Curran maps that arc can be found in episode 'The Cord and the Ax' (season two, episode three), in a scene when Datak, currently imprisoned, is unexpectedly visited by his teenage son Alak Tarr (Jesse Rath), and they discuss the running of the family business. Positioned at an upright rectangular opening in the wall and costumed in white body make-up, a wig and coloured contact lenses, Curran displays a weariness in his character during the opening shots of the scene, pausing frequently to gather his strength to respond, and only gradually building his character's energy across the subsequent shots. A less experienced or not so skilful actor might have either started at, or moved too quickly into, too high a level of energy, leaving them 'nowhere to go' and skewing the dramatic development of the scene. Gradually building energy levels is a particular challenge for screen actors, as they need to ensure continuity and coherence in their performance within a given sequence. Curran gradually raises his character's energy during the shots, as Datak becomes agitated to hear of his son's wish to withdraw from running the family business.

As part of this process, Curran uses a sibilant delivery for Datak's response: 'Out of the queStion, there muSt be a Tarr at the head of the table!' This sibilant quality links phonetically to Castithan, the guttural, rasping alien language formulated for the show, and can also be found, for example, in the vocal performance of Datak's wife Stahma Tarr as played by fellow British actor Jaime Murray. What distinguishes Curran's approach to his vocal performance is that he develops the sibilant quality of his character's voice to a pronounced hiss for moments of particular dramatic significance. This is the case in this scene, when Datak begins to suspect that his wife may be plotting against him. Curran signals Datak's increasing agitation through moving his body and head purposefully with less control, bulging his eyes, and pushing his nose repeatedly through the narrow gap in the wall, prefiguring the forward movement that is about to occur. Datak's anger climaxes when his son angrily rejects Datak's threats towards Stahma and insists that Datak has himself to blame for his current predicament. Curran's angry hiss involves a quick movement of his head temporarily retreating from the narrow gap to allow his right hand to strike forward towards Rath, followed by an aggressive, hissing breath from the back of his throat, with his teeth bared, spit flying from his lips and his nose pushing through the gap (see Figure 9.3). This aurally and visually

Figure 9.3   Tony Curran's Datak Tarr shows his feral side in *Defiance* (Syfy, 2013–15)

evokes the strike of a snake, a feral animal trapped and ready to attack when threatened or provoked.

Curran's decision to draw inspiration from animalistic properties is certainly an established approach in actor training. As Vanessa Ewan and Debbie Green note, making links to animals can be a useful 'staging post' in the difficult process of characterisation, because animals not only have limited and clearly defined qualities, thus demanding 'consistency on the part of the actor' (2015: 127), but also live in the present tense, 'an essential skill for the actor' (2015: 128). Related exercises can be found in many acting text books.[8] For example, James Penrod instructs the reader in his chapter titled 'Developing the Role Through Movement' as follows:

> Another approach to characterization is to use animal imagery. While reading your playscript, think of an animal that might behave in a manner similar to that of your character. [...] If you want to use the chicken as a point of departure for gestural and movement patterns, first try to mimic as precisely as possible the movements and apparent attitudes of the chicken, so as to get these rhythms 'in your body.' Then you can try to simulate the hen's movements *as a human being might make them.*

> (1974: 147–48; original emphasis)

Usually, such an approach is a step within the preparation and rehearsal process, one that can help actors develop their imagination, range of physical movements and engagement with their usual sensory perception. As Ewan and Green have put it in their movement handbook for actors: 'The animal itself presents the actor with inspiring choices about the way a character operates in the world: the motivation for movement, the rhythm of life and the way it sees the world can all be taken into account' (2015: 159). Of course, this methodological tool can and does also feature in finished performance, such as the work here by Curran, who draws on serpentine properties to enrich his characterisation of Datak Tarr.

As with his work for *Doctor Who*, Curran is concerned to use his careful preparation to build a system for his performance: the animalistic hissing is not a convenient one-off, but located within a motivated pattern; only here this has a more pronounced diachronic dimension, as Curran develops the system across the three seasons of *Defiance*. He first uses the snake-like hissing in season one, episode nine and then subsequently deploys it to map key points in his character's narrative trajectory. One of these is the confrontation with his son, in which his acting choices signal that, his character's aspirations towards upward mobility notwithstanding, Curran views Datak as closely linked to the instinctual and primal. His decision to produce such a pronounced hiss – the strongest deployment of the system across the three seasons – also demonstrates and reinforces his character's strong concern with disloyalty by his family and the threat of emasculation.[9] What merits particular mention is that Curran uses the serpentine hissing quite sparsely: it occurs only a handful of times across the three seasons, at judiciously chosen moments. Curran has the confidence, discipline and judgement to place firm boundaries on his use of such animal-related, attention-drawing physical movements and sounds. Indeed, the effectiveness of Curran's systematic approach, both here and in his other roles, partly depends on him balancing the particulars of the system against his overall concern to (appear to) 'do little'.

Roots: Handling Violence, Repetition and Rhythm

Following the cancellation of *Defiance*, Curran successfully auditioned for the final case study in this chapter, namely the part of Connelly, a cruel slave plantation overseer in the recent remake of *Roots*. This is another role that presented him with some particular acting challenges, and Curran's meeting of these relied upon his ability to combine the physical demands of the scene with aesthetic concerns, including his attention towards rhythmicality. Our focus lies with the scene towards the end of Part 1 when Curran's character viciously whips Kunta Kinte (Malachi Kirby) following his attempt to escape, well remembered from the 1977 iteration. On-screen, the scene in total runs for roughly five minutes, containing a multitude of shots. The script, however, was left open so the actors had the freedom to develop the scene on their own terms, drawing partially on improvisation, and it took two days to film. Despite the fact that the violence is of course simulated (aided by some quick cutting), the filming of the scene was demanding for those involved. While press and audience responses to this scene and similar projects, such as *12 Years a Slave* (dir. Steve McQueen 2013), have quite understandably been more concerned with the performance/experience of the actors playing the victim, it is important to stress the demands this kind of scene places on the actors cast for the role of perpetrator. Curran found playing Connelly to be 'one of the most challenging things I've done. I think *Roots* was just psychologically and emotionally very intense, [...] pretty visceral'.

As with the case study from *Defiance*, Curran builds the intensity within the scene gradually. Before the whipping commences, Curran holds himself still in both wider and tighter framing, even avoiding blinking within close-ups. Once the flogging begins, his voice remains quite calm as his character orders Kunta to accept his slave name, Toby. This restraint by Curran indicates not only that this cruel act is a regular occurrence for his character, but also that he expects the situation to be resolved quickly. Curran shifts his energy after ordering Kunta Kinte to 'Say it' for the first time, as Curran has Connelly's exertion and outrage build in tandem. The exertion is only partly performed: as Connelly starts to stagger and sway with jerky movements, and gets somewhat out of breath, so Curran has to use the whip. Benefiting from his long-standing emphasis on maintaining his physical fitness, Curran carefully practised handling the whip in preparation for the role. One of the challenges of simulated on-screen violence is to deal technically with hitting the mark (here, where the whip lands), while delivering an emotionally truthful performance. The emotional truth Curran is here concerned with lies in the growing outrage his character is experiencing (see Figure 9.4), unsettled by the unexpected level of resistance by Kunta Kinte, who refuses to accept

Figure 9.4 Tony Curran's Connelly administers a whipping in *Roots* (History Channel, 2016)

his slave name. That Curran develops this emotional response gradually links to the performativity of the act of whipping, which Connelly delivers in front of the plantation owners and other slaves, and which Curran visually highlights by pointing at the watching slaves. The building of the outrage by Curran signals that Connelly becomes aware that he is not delivering a successful performance as overseer and that his status (both with regard to his bosses and the slaves) comes under threat, which would mean an end to his career at this plantation at the very least.

An additional (and carefully rehearsed) thread that Curran weaves into his performance here concerns the use of repetition in the script, which has Connelly order Kunta Kinte to 'Say your name', 'Say it', and demand: 'Toby's your name. What is your name?' Having already demonstrated his skill at prosody in *Defiance*, Curran here brings variation in terms of pitch and delivery to his vocal performance, and uses the whipping to punctuate his lines, thus infusing the scene with a distinct rhythm. His breath control merits particular attention, for Curran's diction remains precise throughout, even when becoming slightly breathless from the physical exertion. Of course, it is important to be mindful of Pamela Robertson Wojcik's argument concerning the significance of sound technology for screen acting, that 'the voice must never be considered as emanating in any simple way from its visual source' (2006: 75). As she points out, vocal performance is deeply intertwined with sound technology (in production, post-production and exhibition), in such a way that manipulation and design are effaced and disavowed. Curran's work is no exception, and his performance here needs to be understood as 'existing within, for, and through mediation' (Robertson Wojcik 2006: 78) within the show's sound design. However, just as it is important not to efface the role of technology, so it is crucial to sufficiently recognise the significance of actorly craft, skill and labour within the nexus of vocal performance and sound design. If Curran delivered his vocal performance (mostly) on-set, then he contended with the physical effort of using the whip without loss of diction; if ADR (additional dialogue recording) was involved, then Curran had to act being physically exerted while keeping the diction crisp during post-production. Either way, Curran here displays technical and creative skill, bringing together the physical challenge with emotional truth and a strong rhythmic quality across the cuts in the sequence.

Conclusion

Across the course of his career thus far, Tony Curran has played a range of very different roles with particular challenges: some restrained (*This Life*); some much more expressive (*Doctor Who*, *Defiance*); some marked by specific genre demands (*Defiance*), production needs (*Doctor Who*) or physical and/or emotional challenges (*Doctor Who*, *Roots*); and some privileging the needs of the actors over the production methods (*This Life*, *Roots*). Curran's approach to acting has not changed during the course of his professional experience; rather, there is a somewhat cyclical process whereby, following his positive experience of drama school, Curran's confidence and skills have helped him get cast for such diverse parts, and in turn his experience of such a variety of roles has helped him maintain and develop his skills and confidence. As this chapter has demonstrated, Curran's versatility is closely linked to his confidence and thus his willingness to take creative risks, his discipline and emphasis on careful preparation, especially as far as the development of systems is concerned, his skill in combining technical and practical specifics with aesthetic concerns, and his enduring awareness of the significance of the methods of production and exhibition. With these qualities making him a particularly capable and well-rounded performer, his versatility is furthermore informed by the positive, productive working relationships he has with both his long-term agent in Britain and his agent and manager in the USA, as well as the reputation he has built within the industry's social culture for being not only a good actor but also a good actor to work with.

Some pertinent ideas and critical issues have emerged over the course of our analysis in this chapter that deserve to be taken forward in more depth in further discussions of acting and performance. Two strike us as particularly noteworthy: first of all, there are the emotions that actors experience, as highlighted by Curran in the interview. The emotional labour that actors undertake in tandem with their performance work should be given more serious consideration, as the latter is inseparable from the former. And secondly, it is our contention that the burgeoning field of writing about television acting will benefit from further diachronic studies of individual professional actors' work. These will help to further move discussions of acting beyond the prism of stardom, and locate the work of those professional actors within the specific stages of their career against which achievements ought to be mapped (e.g. some acting choices become more noteworthy when delivered by emerging actors). Such studies will bear out the ways in which actors' careers develop chronologically but not linearly, and allow us to see actors' agency as both facilitated and dependent (for example, Curran got an early break partly via the help of a friend, but later the high-profile US show in which he was a lead was cancelled). Most significantly perhaps, they will uncover the richness of actors' work across different projects and contexts.

Acknowledgement

We thank Tony Curran for generously providing his time and thoughts.

Notes

1 For further discussions of Stanislavski, realism and naturalism, see Gordon (2006), Carnicke (2010) and Hewett (2015).
2 See also Moore (2004). This links to issues concerning status, which Roberta Pearson (2010) has discussed in some detail and to which this chapter will briefly return.

3   Self-consciousness (which is to be distinguished from self-awareness) has long been iden-
    tified as a hindrance to (good) acting, if not a signifier of bad acting, in both actor and
    actor training discourses.

4   In addition to those already mentioned, these further include, for example, *Medium*
    (NBC/CBS, 2005–11) and *Elementary* (CBS, 2012–).

5   For further details, see Knox (2018).

6   Lee Strasberg has defined affective memory as 'the basic material for reliving on the stage,
    and therefore the creation of real experience on the stage. What the actor repeats in per-
    formance is not just the words and movements he [sic] practiced in rehearsal, but the
    memory of emotion through the journey of thought and sensation' (1987: 113).

7   Time in combination with another determinant of television acting identified by Pearson
    (2010: 169–73), namely status, afforded Curran the experience of the *Defiance* writers'
    increasing attention to the main cast influencing the scripts. For example, Curran's
    Scottishness is playfully referenced in his character's use of the words 'wee one' when
    dealing with his grandchild in the third season.

8   Ewan and Green advise against choosing snakes for a first study, as these 'are limbless (the
    actor must invent too much)' (2015: 128).

9   Curran was re-watching *The Sopranos* (HBO, 1999–2007) when preparing for this role.

10

The Same, but Different: Adjustment and Accumulation in Television Performance

Lucy Fife Donaldson

Television offers a set of opportunities through performance, both for the performer and the audience. For the actor, serial drama – a term I am using as a broad description of a series with an ongoing narrative developed over several seasons[1] – in particular presents an opportunity to inhabit a character for a far greater length of time than on film, possibly over several years, therefore giving them time to develop and hone their performance. As a result, television can afford the performer a model of acting practice that has many benefits, including the time and space for professional development and stability. For the audience, television permits a familiarity with performers and their gestures, ways of speaking and moving. One of the pleasures of the long-running series is to be found in such familiarity and in-depth knowledge or understanding of a performer's work. Through close attention to one performer and their work across two drama series, this chapter will explore the particular possibilities granted performance by the serial nature of television drama. In doing so it seeks to highlight both the expressive achievement of performative textuality and the evaluative challenges it brings.

Timothy Olyphant is a US actor who has worked in both film and television. In film, he has never regularly progressed to lead roles[2] and has gained more work as supporting characters; his most high-profile film role to date has been the villain in *Live Free or Die Hard* (dir. Len Wiseman, 2007). Television has brought Olyphant more sustained success, and it is here that he has received acclaim in two major drama series:[3] as Seth Bullock in *Deadwood* (HBO, 2004–06) and Raylan Givens in *Justified* (FX, 2010–15).[4] In both *Deadwood* and *Justified* Olyphant plays a law man (a town sheriff and US

Marshal respectively), and although the series are set in different time periods (the 1890s and present day) and geographical regions (South Dakota and Kentucky), there are strong similarities between the characters, almost what we might call a family resemblance (something that *Justified* certainly draws on as a series). The strong likeness in Olyphant's performances forges these connections further, as he uses parallel gestures, facial and vocal expressions, and even ways of walking in both series. Looking across the programmes offers an opportunity to consider these similarities, and especially how his performances build in each, the repetitions across the series leading to accumulations of meaning in gesture, posture and voice, which intricately work with television's 'dense accumulation of citational and narrational history upon which performances may come to draw' (Logan 2015: 28). In part, then, this chapter suggests that to consider both series in relation to each other is to be able to appreciate a layered performance, our understanding of Olyphant's depiction of Givens building on our possible knowledge of his portrayal of Bullock. Moreover, that the medium of television rewards certain kinds of performances, especially the type of detailed accumulative work that Olyphant is engaged in, directs my argument to consider the value of a particular performance style suited to longevity in television drama. Olyphant is chosen for having, on the one hand, a career that is typical of that of many television performers who move from one series to another often within very similar genres or types, while, on the other, offering a distinctive example of a performer in a leading role whose work across series is very similar. Olyphant's singularity is further tied up in his success as a TV actor, as compared to the mixed results of his film work; while there may be many reasons for this (both professional and personal), it might point to there being something about his acting style that suits the medium of television. My choice of case study therefore seeks to highlight the value of a restricted or small acting range as particularly suited to the requirements of serial drama, in counterpoint to the emphasis placed on achievement through transformation that so often dominates popular appreciations of performance.

Moments of Comparison

In order to start the process of articulating the resemblances between Olyphant's performances, the chapter will begin by examining two moments early in each series, taken as broadly typical of both performance and character action, while at the same time highlighting certain performative choices, and the repetitions of these choices, made across the series.

In the final episode of *Deadwood*'s first season Seth Bullock announces his intention to take on the role of sheriff to Al Swearengen (Ian McShane), owner of the Gem Saloon and the strategic lynchpin of the town. The antagonism and leery partnership between Bullock and Swearengen is one of the central dynamics of *Deadwood*, and the encounter is one of many exchanges they have during the course of the narrative's events. This instance is remarkable for being a point where Bullock accepts Swearengen's leadership and recognises that he is acting in the camp's best interests, despite any personal misgivings he holds about him. Bullock's decision is made in the context of the imminent departure of General Crook (Peter Coyote) and his troops, along with the man who Bullock beat nearly to death earlier in the episode, Otis Russell (William Russ), who is also the father of Alma Garret (Molly Parker), a wealthy widow with whom Bullock has just begun an affair. Bullock and Swearengen's opening discussion about Crook and Russell is cut short by Bullock announcing that 'if that man [Russell] comes back to the camp he'd be my problem to deal with [...] I'll be the fucking sheriff'. Al asks him to put the badge on, which Bullock does somewhat reluctantly, and they drink.

Olyphant's performance throughout the scene can be characterised by containment, especially in comparison to McShane, whose performance across the series matches Swearengen's frequent performatively rhetorical flourishes, using larger gestures and greater modulation in his voice; Steven Peacock refers to McShane as occupying a 'supercharged expressive register' (2010: 109). In contrast, Olyphant maintains his reactions within a smaller emotive spectrum. Upon entering the room above the saloon, he comments 'there's a bloodstain on your floor', but moves past it without stopping, his voice almost monotone as he delivers the line. Swearengen's deflection of his observation 'Yeah ... I'm, er, gonna get to that' draws only a slight nod and eye/eyebrow flick of acknowledgement. At the end of their short discussion, they raise their drinks, with Olyphant looking sideways for a beat before clinking glasses, this eye movement the sole external indication of Bullock's misgivings concerning the decision he has made and this new strategic alliance with Swearengen.

The conversation progresses as a series of terse exchanges, with Swearengen behaving more expansively while Bullock remains unwilling to talk. His increasing frustration and annoyance is demonstrated through a series of controlled and taut gestures: he stares forwards intensely at Swearengen for most of the scene, his hands clenched by his sides as he sits opposite him. When Swearengen reminds Bullock of his indication to Swearengen's hench-man Dan Dority (W. Earl Brown) that Alma's father would be better off dead, Olyphant takes off his hat and briefly holds the bridge of his nose, then looks up and gestures towards McShane with his left hand held flat, his eye-brows raised and jaw clenched as he stresses his point (Figure 10.1).

The tension culminates with him standing to produce the sheriff's badge, and when Swearengen prompts him to wear it – 'on the tit' – Olyphant further tightens his jaw, his frame inclined forward, the clenching spreading through his body as he grimaces, saying in a low voice 'I know where it goes'.

Moving to *Justified*, a broadly comparable scene occurs at close of the first episode which culminates in a showdown between Raylan and Boyd Crowder (Walton Goggins), the character who will go on to be his primary antagonist and counterpart throughout the series. In certain respects, their relationship matches the one between Bullock and Swearengen; they share a similar dynamic of terse (Bullock/Raylan) and verbose (Swearengen/Boyd), and Goggins' performance, like McShane's, is bigger, with larger vocal

Figure 10.1 Olyphant looks directly at McShane and uses his left hand to emphasise Bullock's resolution on Russell (*Deadwood*, HBO, 2004–06)

modulations and rhetorical/gestural embellishments. Over dinner at Ava Crowder's (Joelle Carter) – former sister-in-law to Boyd and initial love interest to Raylan – their discussion ends with Raylan shooting Boyd in the chest.

Although Raylan is somewhat more relaxed and willing to speak than Bullock, Olyphant deploys related gestures to communicate his focus and tension, becoming increasingly restricted and still as the scene goes on. Once he has taken off his hat, he continues to hold his head at a slight incline, looking up as though the hat is still in place, a posture that echoes the one adopted in *Deadwood* and forces him to keep his eyebrows raised and forehead wrinkled. Throughout their conversation Olyphant holds eye contact on Goggins. This is maintained more loosely at the start with some eye movements between him and the table, but as the confrontation intensifies and Boyd forces Raylan into a shoot-out, Olyphant stares straight ahead at Goggins as he did with McShane. Even when Ava interrupts, training her shotgun on Boyd, Olyphant only looks at Carter briefly (an edit replaying the shift and prolonging it) and continues to watch Goggins. Olyphant employs the same hand gesture – flattened left palm held in front of him, perpendicular to the table – to underscore an emphatic statement: 'I thought I had till noon' (Figure 10.2).

Figure 10.2 Olyphant repeats the combination of forward gaze and gesture with his left hand, with a slight modulation of force in *Justified* (FX, 2010–15)

At one point Olyphant shifts out of the taut range of postures, gestures and vocal inflections as Raylan attempts to deflect Boyd's threats, by lightening his closed expression. He moves his head, looks down and then up, raises his eyebrows and smiles as he says 'you can call it off', his voice rising more expressively than before, an expulsion of breath audible as he smiles on 'off'. Yet this modification is only brief; almost immediately afterwards his face hardens and in the next beat his expression and voice close down, becoming tense and tight as Raylan stands to shoot Boyd – following Ava's intervention in their showdown – clenching his jaw and momentarily frowning as he stares at the other man.

These two short examples provide a basis from which to compare Olyphant's performances in each series, developed through a narrow set of controlled components: slight eye movement, consistent eye contact, inclined head posture, direct hand gesticulation, clenched jaw and flattened line delivery. Restraint is the primary register, the characters requiring only minimal shifts in tempo, energy and tone.[5] This quality is recognised by both Steven Peacock and Jason Jacobs in their writings on *Deadwood*. Peacock praises Olyphant, along with Molly Parker, in particular for the way they 'encourage a restraint of the kind of broad-brush actorly tics and mannerisms that normally pass for dramatic characterisation on television' (2010: 108), while Jacobs describes Olyphant's *Deadwood* performance as 'never less than intensely quietly brilliant' (2012: 94). In his writing on *Justified* William Rothman alludes to this restraint, describing Olyphant's characterisation as in-depth rather than on the surface, the lack of a visible expressivity giving Raylan psychological complexity: 'In the pilot, Timothy Olyphant's Raylan [...] incarnates, in the way every real human being does, the mystery of human identity [...] an unfathomable depth, a dimension of unknowness, that cannot simply be the invention of a writer' (2013: 178). The descriptions of these writers could in each case apply to Olyphant's performance in the other series.

A further result of this comparison is to understand Olyphant as working within a limited range – utilising the same or only marginally modulated gestures, postures and tones – and that his performances are therefore more generally not geared to make large distinctions between the characters, or to be particularly recognisable as acting as defined by traditional notions. For Andrew Klevan '[s]uccessful acting is often wrongly, or misleadingly, understood in terms of range' (2003: 18), a criterion for film performance in particular that leads to having a small range mobilised as a criticism of an actor, the complaint that they are always the same, not sufficiently versatile. Klevan finds there to be no such correlation in his appreciation of Joan Bennett, suggesting that her 'limited repertoire of characteristics [are] nevertheless capable of developing an expressive rapport, and relevance with the other elements of a film's style, and require only slight adjustment to adapt suitably to changing situations and differing moods' (2003: 18). Likewise, it is the aim of this chapter to argue that such narrowness is something to be appreciated as part of a successful television performance, and that the characters acquire considerable complexity through the subtle differences and changes adapted for use with each. What's more, through the accumulation of time spent in a role facilitated by serial drama, the performer is able to intensify their expressivity within that small range.

Performance Tools

While the two moments described above are valuable for fleshing out some detail of Olyphant's performances, they can't capture the wider sense of how these performances work, on their own or together. A major challenge in the work of attending to television performance, especially a comparison involving two drama series totalling 114 episodes, is that an individual moment, no matter how exemplary or remarkable, doesn't engage with the scope or overarching texture of a performer's work. Yet performance analysis in any medium requires the depth an investigation of a moment provides. Although writing about film performance, Klevan's understanding that close attention to performance is not just necessary but prompted by the expressive capacities of the performer and the drama certainly applies to television:

> The close study of films and their performances, therefore, is not simply a particular method of film analysis, personally favoured; it is invited by the 'variations' and 'permutations' of the drama, by the 'intricacies' and 'richness' of the performance, and by the 'repeatability' of the medium.
>
> (Klevan 2005: 7)[6]

This is perhaps even more true given the considerably greater depths of variation and intricacy afforded by the span of serial drama. Indeed, engagement with acting on television highlights that there is a larger and thicker scope available for performance in this medium – built through time spent (and the possible challenge for the actor as recounted by Roberta Pearson (2010)), which allows for detailed work by the actor and immersion in detail by the audience, as well as repetition and familiarity with the character for both actor and audience.

Building on our engagement with the moments above, then, we can respond to this analytical challenge by identifying the primary facets of Olyphant's performances as Bullock and Raylan, in order to (1) expand understanding of what constitutes the parameters of his range, and (2) attend to the density of repetition, and, perhaps more importantly, every adjustment.

- **Movement**. The principal component of Olyphant's movement is his walk, in which he carries a distinctive stiffness. Peacock offers an evocative account of Bullock, through which he stresses the generic traces of his gait through the western: 'Olyphant walks with the wary purpose of a lawman in unruly terrain, channelling [Henry] Fonda's lean, stiff grace; his eyes have a feline flash and his legs a nimble gait, yet he always swivels his neck and walks as if carrying a book on his head, always *poised*' (2010: 106). This poise combines the underlying tension between the natural ease of his lithe muscularity and the containment required to restrict the violence he is capable of.
- **Stance**. In stasis, Olyphant continues to be poised, organising his body into a series of postures that combine stiffness and ease. Frequently he stands with one knee bent, giving the appearance of a more relaxed and less formally upright position while the bent leg imbues it with a degree of rigidity. In *Justified* he expands his still posture to leaning (perhaps on a door jamb or a banister), though frequently with arms crossed over his chest. The lean manages to be both flexible and stiffly closed at the same time (Figure 10.3).
- **Gesture**. The most consistently repeated gesture is that of his hand held out flat to emphasise a statement. The angle at which the hand is held – straight out or at a slight upwards incline – has purpose in slight variance, indicating degrees of openness or resolve. Olyphant also rubs the bridge of nose in gestures that express shades of exasperation, annoyance and frustration.
- **Face**. As befits the reticence of the characters, Olyphant is not prone to revealing much in his facial expressions, limiting these to a series of small

movements that can be contained within one portion of his face at a time (small adjustments of mouth or eyes). Jacobs understands this as a masking: 'A silent exterior masks, and in its stillness, implies an unspecified energy and power which dominates the spaces around it and confuses those who interact with him; he is not skilled in deceit, not good at hiding the emotions he projects so powerfully' (2012: 56–57). This masking means that Olyphant's face is typically concerned with maintaining composure and enacting compression, in efforts to stifle his characters' rage and frustration.

- **Eyes**. Prolonged eye contact, or at least a direct and unmoving look (towards another character or down), is characteristic of the way Olyphant uses his eyes and eyeline. As a result, even the smallest change to the direction of his gaze or his face around it is imbued with meaning. As at the end of the *Deadwood* scene described above, he employs a flick of his eyes to the side (or down) to communicate moments of doubt or regret. The eyes move frequently in combination with mouth movement, a firmly set mouth compressing the eye flick into a harder expression.

Figure 10.3 Related ways of standing: left, in *Deadwood* (HBO, 2004–06); right, in *Justified* (FX, 2010–15)

- **Mouth**. Although both Bullock and Raylan do smile occasionally (Raylan more often), Olyphant tends to hold his jaw in a taut position, clenching it further at key moments (often combined with lowered eyebrows or a furrowed brow). This has significant impact on his speech patterns as it produces degrees of flattened and lowered vocal delivery, forcing the voice into the very front of his mouth and then his throat the more his jaw is tightened. Another related mouth movement is to purse his lips, when frustration or anger overtake his typically still exterior. Smiling signals shades of ease and politeness.
- **Props/costume**. Both characters share similar costume and props, most notably the Stetson-style hat and gun with holster belt. Action enables a range of opportunities to interact with these items, which also add to some of the positions and movements already described. There are three key uses of these items: placing the right hand on the holstered gun/belt (leaving the left free to gesticulate); taking off/putting on the hat in gestures of politeness and matter-of-factness; looking up from under the hat (whether it's on or not) which often causes him to tilt his face down while raising his eyes.

This taxonomy of gestures, postures and mannerisms are repeated across both performances and while they are incorporated elsewhere to a certain extent in Olyphant's other performances in both film and television, the repetitions between *Deadwood* and *Justified* are contained more precisely within this core range. Identifying these – what James Naremore (1988) would refer to as his idiolect (the difference being here that these traits are not part of a star identity) – provides a baseline of expressivity from which to create a detailed map of how the performer signals characterisation and response to the fictional worlds of Harlan and Deadwood.

Taken across *Deadwood* and *Justified*, we can further build that map into a highly detailed chart, becoming more and more capable of discerning the meaningfulness of the tiniest instances of these performative elements and their slight adjustments (the angle of the hand, the extent to which the leg is bent, how long the eyes flick away for). In their study of Lesley Sharp, David Forrest and Beth Johnson adopt an approach that considers how her northern identity is developed intertextually through multiple televisual performances, arguing that her 'presence and performances are in many ways spectral, each haunted by what has gone before so that every new performance bears traces of the past' (2016b: 201). While the intertextual layering across Olyphant's performances in these two series is arguably more materially present than a haunting trace, and thus perhaps related to distinctions between star identity (Sharp) and performance (Olyphant),[7] it is nonetheless revealing that Forrest and Johnson conclude by praising Sharp's 'extraordinarily subtle performances of gesture, voice and movement' (2016b: 201), their attention equally captivated by restrained expressivity succeeding on television. As Forrest and Johnson find with Sharp, attention to Olyphant's set of gestures presents not a reduction of expressivity, but rather a minutely calibrated sense of recognition of the character's behaviour and how meaning might shift from moment to moment, or series to series.

Moreover, a viewer familiar with *Deadwood* might come to *Justified* better equipped to apprehend the specificities of what Olyphant as Raylan is doing. The repetitions especially suit a performance that remains contained; the tendency towards stillness and flattened expressivity is mirrored by the reduced range of postures, gestures and movements. The compactness of his performative range is thus determined by certain emotional similarities of the characters – especially their anger, as we will explore below – which present opportunities for related expressivity. Repetition is further driven by the fact that the characters are involved in parallel activities, repeated actions and pieces of business arising from clashes between the law and the criminal, such as standing to draw a gun, whether dictated by their professional capacities as lawmen or their personal impulses, which run between a gamut of violence and politeness. Generic resemblance plays out in shared iconographies of costuming and props, which entail a look from under a hat, walking and putting a hat on, handling a badge and gun.

This is partly, then, a consideration of the relationships between genre and performance, as both series are westerns of one sort or another (or, perhaps more accurately for *Justified*, Raylan adds a western element to what otherwise might be regarded as a form of police procedural drama) and generic resemblance is played out in shared iconographies of costuming and props, and their related actions and gestures. When Peacock refers to Bullock's 'lockjaw physical terseness' (2010: 111), the phrase holds a connection to depictions of the western genre's tropes of white masculinity in both film and television (as Peacock and Jacobs observe, Olyphant's Bullock carries shades of genre-defining film performances by Henry Fonda and

Clint Eastwood), and can equally apply to Raylan, though there are grades of distinction still to be made between the two. Indeed, applying Klevan's ideas on range to a male performer might further help us reflect on how such restriction can be a hallmark of certain kinds of masculinity, especially those tied to the western, where consistency and a curbing of emotion are seen as strong and valued masculine traits. For Jacobs, this generic history 'is reinforced visually by a repeated shot of Bullock walking with determined purpose through Deadwood, while the rest of the busy camp continues around him – he is, in this striding mode, hostage to his own purposes, drives and interests rather than those of the camp' (2012: 56). In *Deadwood*, the western genre is contained and perpetuated through the specificities of his movement and how these position him, as well as his interior life as the walk is both communicative – as observed by Jacobs – and part of the construction of a silent, at times blank or automaton-like, exterior. *Justified* embellishes this western inheritance, weaving it more frequently into pieces of action – as with the armed stand-off enacted periodically throughout the series' narratives, from the opening scene of the first episode to the denouement of the last – to build Raylan's inhabitation of that laconic masculinity and puncture it periodically, as in a deft exchange with his colleague Rachel (Erica Tazel) in which she points out the privileges he enjoys as a 'tall, good-looking white man with a shitload of swagger'. In a moment like this, *Justified* draws attention to Raylan's fit with the world around him and the relative ease with which he can manoeuvre through it, and the corresponding generic appropriateness of the lawman embodied by a white male body, one that is athletic and handsome, violently expressive and emotionally withdrawn.

Olyphant's selection of performance tools allows for the similarities of the characters, and the programmes themselves, accommodating and building a repertoire that condenses and consolidates over time spent as Bullock and Raylan. Yet to recognise the parallels is to become increasingly aware of the distinctions, held in the small elaborations of his articulation of a posture or motion. The precision of his performance allows for the slightest change in energy, plasticity or combination of expressions to simultaneously draw the characters further apart and pull them closer together. In what remains, the chapter will take two areas of significant overlap between the characters' attitudes and emotional dispositions to push further into the layering of similarities and differences.

Containment: Reluctance and Resignation

The restrained register of Olyphant's performances in both *Deadwood* and *Justified* is calibrated in part to suit the characters' disposition of reluctance and/or resignation. Having given up the job of Marshal in Montana and moved to a settlement initially outside the Union and unencumbered by law, Bullock is disinclined to become re-entangled with such a position (at first taking the position of health supervisor to avoid becoming sheriff) and seeks to live more freely as a shopkeeper. He is hesitant at becoming involved in other people's business or the town's workings at all, but at the same time he cannot avoid taking up a position of moral rectitude – in season one he leads a mission to save an orphaned girl and hunt her killer, takes on Alma's business interests in the wake of her husband's death as a favour to Wild Bill Hickok (Keith Carradine) and continues to become further embroiled as the series progresses. Raylan is forced to return to his roots in Harlan County, Kentucky and the criminal landscape of his past (including his father and peers, such as Boyd) following the shooting of a criminal in Miami. Raylan's local knowledge is of great advantage to him throughout the series, though he remains reluctant to fully reintegrate himself and is at odds with both the criminal element, with whom he shares history and kinship, and his law enforcement colleagues, who make it clear how difficult he is to work with. As befits their generic heritage, both characters live by a moral code of honour and duty, which entails operating within a rigid set of behaviours. As a result they choose to keep their emotions contained, as though such expressive resistance will prevent involvement.

Olyphant's performance choices secure this foundation of emotional resistance, reluctance to engage and resignation when the turn of events makes it necessary to act. Both Bullock and Raylan remain closed through patterns of posture and motion; Olyphant's eyebrows and mouth are held in a neutral or horizontal position, thereby reducing the display of emotion, while at the same time making any contraction or release more noticeable. This horizontal alignment emanates from holding his jaw in a set position, which keeps his mouth and eyes steady while working to flatten his voice. Bullock embodies the more extreme version of this rigidity, frequently remaining still and wordless in response to events around him. At the moment when he and Alma are about to consummate their flirtation, Bullock shuts the door and for a couple of beats he stands facing it, with his back to her; Olyphant's hands hang by his sides and his head remains upright and stiff. Bullock's facial hair – a trimmed moustache and goatee – further compresses his mouth into a straight line and the overall stillness of his face means that his voice remains in a low monotone register. For Raylan, the clenched jaw is loosened a little: Olyphant's chin sometimes moves from side to side as he listens to a colleague or antagonist, a more relaxed position that allows for a greater inflection in his line delivery. Raylan also has a softer stare than Bullock, the more substantial eyebrow flicks and creases around Olyphant's eyes accommodating the character's wry attitude, itself another form of deflection or resistance to emotional entanglement.

There are motions of compression adopted by Olyphant that serve to turn the characters' energies further inward. This is perhaps best encapsulated in the gesture he uses in the *Deadwood* sequence discussed above, when he brings his hand to the bridge of his nose, leaning forwards into the action as he does so. Using the same gesture in *Justified* during a conversation with his former boss Art Mullen (Nick Searcy) about catching Boyd, Olyphant leans back and draws his fingers over his eyelids towards the bridge of his nose. For Bullock the gesture is used to push down his frustration before a quick movement forward to deliver his intentions to Swearengen, whereas for Raylan the prolonging of this motion with his hand, combined with a reclined position, seems an attempt to slide the frustration away, thus dissipating the energy associated with it.

A key piece of costume, the hat, is deployed to contain emotion. As already mentioned, Olyphant frequently adopts a position of looking out from under the hat, even when not wearing it. While this decision might seem to force him to adopt a more expressive countenance, as it causes him to raise his eyebrows in a more dynamic expression than his typical stillness and so forces his features into the appearance of greater engagement with the world around him, keeping the hat on enables him to keep his face guarded. The action of putting the hat on further contains any emotional impulses, just as taking it off might release them, opening access to his face and drawing his arm in an expansive outward movement. In preparing to remove a competing salesman from their pitch, Bullock carefully places his hat on his head and moves forwards, the motion hardening Olyphant's features into their set flattened position. For Raylan, the hat becomes even more of a guard, often to the frustration of others who might be trying to read his motivations, with Olyphant frequently keeping his stance tipped forward enough for the hat to cover his eyes (Figure 10.4).

These shared gestures and postures of reticence and containment sharpen the distinctions between Bullock's intense rigidity and need to compress emotional flux into a flat line and Raylan's desire to push away and hold back from emotional engagement. The fractional adjustments of fluidity and movement Olyphant brings into his embodiment of Raylan might allow for a momentary easing of tension but the character's surface is no less hardened and set than Bullock's.

Anger and Force

The accumulations of compression feed into containing the characters' anger, which is frequently expressed through violence; this is more explosive in Bullock's case, prompting Swearengen to quip, 'He's got a mean way of being happy', but typically holds more deadly consequences for Raylan, whose propensity for shooting people highlights an aggression not wholly contained by his employment in the Marshal service. During an exchange that closes the pilot episode, Raylan's ex-wife Winona (Natalie Zea) tells him,

Figure 10.4   Looking out from under his hat in *Justified* (FX, 2010–15)

'you're the angriest person I have ever known'. Both men attempt to escape aspects of their past and as a result face tensions arising from their inability to shake old behaviours, roles and even associates, which are given some form of release through their status as lawmen. As Jacobs notes, for Bullock this outlet is exceeded by the force of his emotions: 'such currents of extreme violence and quick temper mean that the order he could impose through his skills in law enforcement is constantly threatened by more private loyalties and feelings of individuated hurt' (2012: 57). Likewise, however detached he might seem, Raylan is consistently unable to break away from pursuit of Boyd or other elements that threaten Harlan.

Anger thus provides an emotional bedrock for Olyphant's performances. This is held most significantly in the tautness of certain movements and positions and the small expressions – an eye flick or eyebrow wrinkle – emerging through clenched limbs and jaw; anger must be controlled beneath that expressionless surface. The stiffly purposeful gait holds together both the energy afforded by anger and perhaps more importantly its suppression, the repeated views of him walking in *Deadwood* giving the motion a relentless consistency.[8] As Bullock walks through the main street of the camp to confront Otis Russell, Olyphant's eyes are fixed forward, arms held loosely by his sides so they are thrown outwards with the energy of his rhythmically even and propulsive pace. When his partner Sol Star (John Hawkes) tries to intervene, he responds, without any change in gait or gaze, 'Get away from me Sol, get away', in a low tone, hardly moving his jaw and lips. There is a dynamic directness in Olyphant's gait, rolling inexorably towards a target, even when taken at a lesser speed. In the moments leading up to his beating of Russell, while the other man talks Olyphant keeps his gaze still – either looking straight ahead towards William Russ, or, as Bullock's fury increases, positioned down to the floor – and registers his rising temper through jaw movements, slight at first so that his moustache seems to gently ripple, and then more prominent twitches as he clenches his teeth and jaw, causing his nose to slightly flare. When Bullock finally overboils – prompted by Russell ending his speech with 'you'd better take your swing' – Olyphant's face compacts inwards and his punches, thrown with his right hand, come as regularly and relentlessly as his pace through the camp. Other than breathing with the effort, there are no further expressions of rage, his anger directed fully into the force of his right fist, and the tight hold he maintains on Russell's collar with his left.

Although Raylan has more time for leaning, a bearing that could express ease, his movements are likewise purposely direct and performed without hesitation, his violent impulses directed into fluid motions of disarming and hitting, as when he first encounters Dewey Crowe (Damon Herriman) in episode one, and firing a gun, which he does frequently. While Bullock's control is overturned momentarily, as with Russell or during a spectacular fight with Swearengen at the beginning of season two, Raylan's anger is subsumed into the frequent acts of lawful, or justified, violence perpetrated through his role as a Marshal. A series of tense encounters with backyard fighter/wannabe cock-fight promoter Randall Kusik (Robert Baker) culminates in Raylan shooting Randall with a beanbag gun, given to him by Rachel to avoid the inevitable violence leading to anything worse. Raylan resists displaying his anger, walking stiffly from his car to lean with the shotgun resting on the hood, his head held down to cover his eyes with the brim of his hat. Olyphant delivers his opening line – 'that's close enough' – with a cadence descending into a monotone. Where Bullock's eyes are widened as he stares at Russell, Raylan creases his into thin slits as he talks with Randall. When Raylan's first shot is fired, Olyphant uses a direct, quick action to reach for the gun and press the trigger all in one motion. This is not unlike Bullock's first punch, but there remains a significant distance (by virtue of using a gun) as the rest of Olyphant's body remains still. While a shift to a low angle shot dramatises the movement of the gun, it also intensifies Olyphant's stillness. For Raylan's anger, Olyphant's gestures, postures and motions seem to calcify into brittle toughness despite the affecting of an ease played out in his leaning or a hand placed on his hip.

As with the containment of a variety of emotions, anger is further channelled into a series of forceful gestures, eye movements and voice patterns. The flattened palm is a good example of the oft-employed gesticulation that can become more aggressive through Bullock – the hand perpendicular to the ground – or more persuasive/cajoling with Raylan – the hand held with the palm up. The direct gaze and continuous eye contact with an antagonist takes on an aggressive force, Olyphant hardening his eyes through a rigidly composed stare (Bullock) and a squinting frown (Raylan). While as Bullock Olyphant's voice evens and lowers, sinking into the fury of his tightened jaw, as Raylan he might raise his pitch and increase the tempo, as during a confrontation with Ava, now his confidential informant, in season six, Olyphant clipping his words away from the detached and more drawling accent he adopts for the Kentucky native.

Having laid out these interrelated aspects of overlap between the characters, there remain many gaps, especially in the broader distinctions between them. Bullock's quiet, often grimly determined intensity finds a counterpoint in Raylan's more verbose (though only in relative terms – there are many other characters who talk more in *Justified*) and dryly sardonic humour, Olyphant supplying a twinkle in the creasing around his eyes that Bullock rarely shows. In their shared dispositions, expressed through Olyphant's motions of directness and compression there are therefore shades of softness, wryness and intensity. The performance tools outlined above provide a spectrum from which the performer adjusts to more or less rigid, tense, still or contained.

Conclusion

In his writing on performance in *Deadwood*, Steven Peacock adopts Manny Farber's notion of 'termite performances' to account for their being geared towards a focus on the immediate present through engagement with little bits of business, accumulation and withdrawal from anything of a larger register (2010: 110–11). Perhaps unsurprisingly, one of Peacock's prime examples of this tendency is Olyphant, evoking a parallel to Farber's description of Henry Fonda as defensive, modest and stiff. Bringing in *Justified*, this becomes an apt account of Olyphant as a performer more generally, and of his performance choices, which are a way of 'burrowing into', to borrow Farber's phrasing of termite qualities (1971: 11). The restraint characterising his performance in both series is expressed through a set of gestures, movements and vocal expressions that articulate containment, compression, directness and force. In contrast, other male television performers more widely celebrated, such as Bryan Cranston (*Breaking Bad*) and Jon Hamm (*Mad Men*),[9] depend on change and expansion, their performances engaging expressively and thematically with masquerade (Hamm) and transformation (Cranston) and involving levels of anxiety and psychological torment that are commonly read as 'good acting'. The comparison could be opened out further to consider what this reveals about performances of white masculinity more broadly, especially in terms of issues of control and power, for now the limits Olyphant sets contain just as much expression and expressivity, working by condensation and re-articulation rather than growth and enlargement. His reduced range therefore gains an accumulated weight and rich texture as the series progress, connecting moments and giving a consistent shape to the whole. While there is certainly more to say, especially in terms of the compositional strategies of how performance is framed in each series, I hope the analysis within this chapter connects the performances within their own parameters, and at least begins the process of recognising their layered qualities.

The kind of layered textuality of performance described in this chapter is enabled by the medium of television. And while repetition and accumulation are also possible in film and theatre, they are intensified by the expanded time and space of television. While we might take Olyphant as a more extreme example of a performer whose work is able to capitalise on similarities between roles, to carry forward and build on performances that came before, the textuality that I have examined doesn't conform to the identifying echoes created through a cinematic star persona formed through recognisable elements within and beyond the text, nor does it follow the theatrical ghosting of expectations described by Marvin Carlson (2003). Although his performance choices in *Justified* build on those in *Deadwood*, and the character is certainly shaped to play to the strengths of his achievements as Bullock, drawing on the favouring of 'mental and physical attributes' (Carlson 2003: 68), this is not a case of feeding on a stock character or an audience expectation linked to familiarity with his performance style.[10] Rather, Olyphant's performance style is not declarative enough to be emphatically recognisable, nor to build on an outward association between roles. Marshall McLuhan describes television acting as 'extremely intimate', shaped as such because 'the audience participates in the inner life of the TV actor as fully as in the outer life of the movie star' ([1964] 2009: 346). As with Farber, his emphasis is on an inward energy, in part to do with the attention of television to interiority (which for McLuhan is led by the frequency of the close-up), but more precisely because of the particularities of our engagement with television. The repetitions and refinements I have described are noticeable precisely because they are presented through this medium and this format: 'serial television's particular opportunities for involving the audience in the slow accretion of a mutual history between viewer, performers, and characters, achieved through the repeated patterning and minute adjustment of behaviour, attitude, style, and tone' (Logan 2015: 33). As Logan recognises, repeated contact over several years allows the performer's fine-tuning of their style and encourages appreciation of this by the viewer. Moreover, there is a pleasure in noticing this detail for the audience, in being involved and aware of the granularity of acting decisions and their expressive impact. Simply put, television rewards detailed performance and attention to it, and good television performance recognises that.

Acknowledgements

An early version of this chapter was presented at a research seminar at the University of Stirling, and I would like to thank the participants for their discussion and several pertinent questions that have helped shape my approach. I would also like to thank Paul Flaig and my co-editor James Walters for their helpful suggestions.

Notes

1   As defined by writers like Sarah Kozloff (1992), Glen Creeber (2004) and Jason Mittel (2015), among others.

2   The few examples include the eponymous character of *Hitman* (dir. Xavier Gens, 2007), the romantic interest in *Catch and Release* (dir. Susannah Grant, 2007) and the sheriff in a remake of George A. Romero's horror film *The Crazies* (dir. Breck Eisner, 2010), all examples of mid-budget genre cinema.

3   Other television roles include side characters or guest appearances in various comedy and drama series including *Sex and the City* (HBO, 1998–2004), *Damages* (FX, 2007–12), *The Mindy Project* (Fox/Hulu, 2012–), *The Office US* (NBC, 2005–13), *The Grinder* (Fox, 2015–16). More recently he has taken on a co-starring role with Drew Barrymore in the comedy *The Santa Clarita Diet* (Netflix, 2017–).

4   For his work on *Justified* Olyphant was nominated for several awards including Outstanding Lead Actor in a Drama Series at the 2011 Primetime Emmy Awards and a Critics' Choice award for Best Actor in a Drama Series in 2011, 2012, 2013 and 2015. Notably, Olyphant plays a fictionalised version of himself in *The Grinder*, specifically as a successful television actor, a role that indicates the level of visibility he has achieved following *Deadwood* and *Justified* and at the same time cuts across the terse containment of these roles with a more relaxed persona.

5   Although this is not a description that could characterise all his performances, as Olyphant is certainly capable of a much 'bigger' performance style, as evidenced by his roles in *Scream 2* (dir. Wes Craven, 1997), *The Mindy Project* and, most recently, *The Santa Clarita Diet.*

6   In this passage Klevan quotes Charles Affron (1977: 8).
7   Though this is not to suggest that these are mutually exclusive, but rather representative of the different claims each argument is aiming to make.
8   As noted by Jacobs in an image caption, 'Bullock's rage propels him through the camp' (2012: 57).
9   Bryan Cranston was nominated for three Golden Globes, three Primetime Emmys, and three SAG awards, and won one Golden Globe, six Emmys and two SAG awards. Jon Hamm was nominated for four Golden Globes, six SAG awards and seven Primetime Emmys, and won two Golden Globes and one Emmy.
10 *Justified* is not in this sense what Carlson refers to a 'vehicle play' (Carlson 2003: 68), or a star vehicle, the equivalent in film, although Olyphant's role as a producer/executive producer on the show (from season two) does acknowledge a kind of star power.

11

Analysing Aniston: Tonal Complexity and Non-Comedic Approaches to Sitcom Performance

Lydia Buckingham

Instances of poignancy and the use of non-comedic approaches to a line or action are not unusual in contemporary American sitcoms, nor are they restricted to contemporary sitcom. As Steve Neale and Frank Krutnik note in their book *Popular Film and Television Comedy*, the success of the radio and later television sitcom *Amos 'n' Andy* (WMAQ, 1928–43 and CBS, 1951–53) was due to the fact that 'situations developed not for comedy alone, but [also] along the lines of dramatic progression and conclusion' (Neale and Krutnik 1990: 229). While the intention to induce laughter is perhaps 'the genre's primary purpose' (Mills 2005: 8), how actors establish and explore a situation is arguably as important as how they exploit a scene's comic potential. On the whole, non-comedic acting is used more sparingly than comedic but it is far from a foray into uncharacteristic waters. It takes very little viewing of sitcom series to come across a reading of a line or execution of stage business imbued with a sense of pathos which is prevalent enough to form an integral part of the texture of sitcom performance. The non-comedic also extends to entire scenes, examples of which include Jackie's admission that she is the victim of domestic abuse in *Roseanne* (ABC, 1988–97, 2018), Marshall's father's death in *How I Met Your Mother* (CBS, 2005–14) and Raymond's (temporary) failure to wake up after an operation in the final episode of *Everybody Loves Raymond* (CBS, 1996–2005). Non-comedic performance serves an important function in the sitcom. Closing the emotional distance between character and actor that comedic acting arguably exploits,[1] it fosters viewer empathy for the characters. Moments that

209

invite the viewer to feel *with* the character because of their situation rather than laughing *at* them help to build emotional investment in sitcom series. My focus in this chapter is how Jennifer Aniston reconciles the comedic and non-comedic moments in an episode of *Friends* (NBC, 1994–2004), the latter half of which is geared towards conveying the characters' genuine upset in a serious tone. Episode 16 of season three revolves around Rachel's processing of what she sees as Ross's (David Schwimmer) infidelity and her ultimate termination of their relationship, a pivotal narrative development in the series. Attending closely to subtle tonal changes, my analysis celebrates Aniston's complex manipulation of a scene's mood and seeks to nuance neat dichotomies of 'comedic' versus 'naturalist' and 'performance' versus 'acting'.

'The One with the Morning After' was written by the creators of *Friends*, Marta Kauffman and David Crane, and directed by James Burrows, co-creator of *Cheers* (NBC, 1982–93) and veteran sitcom director and producer, who directed the majority of the first season of *Friends*. The reunion of the cast with the original creative team gives the episode a familiar feel which, combined with the sense of sombre finality that Aniston is able to convey at the end of the episode, creates a nostalgic aura that heightens the drama considerably. Since the pilot episode, Ross had courted Rachel. After several near misses and false starts, they eventually maintained a stable union, which began in the last half of season two. Over the course of 64 episodes (and two-and-a-half years, if watched when originally broadcast), Ross and Rachel's relationship has been yearned for, delayed and enjoyed by the characters. The rapid dissolution of it in this single episode, therefore, abruptly changes the trajectory of what has up until this point in the series been a slow-paced 'will they? won't they?' narrative. When I first watched it on Channel 4 in 1997, the end of the episode came as a particular shock. Over 20 years later, after repeated viewings, I find that the sense of loss that Aniston conveys still resonates as unexpectedly heart-breaking. Yet comedic intent also plays an important role in her performance in this episode. There are several moments in which Aniston is clearly playing for the laugh. Taking Susan Berridge, Simone Knox and Gary Cassidy up on their suggestions to conduct further research into 'how you feel and why you feel' when you watch Jennifer Aniston (Berridge 2014; Cassidy and Knox 2015), this chapter explores the relationship between what I identify as Aniston's signals of comedic intent and her postures, gestures and vocal delivery that communicate emotional pain more viscerally.

The episode begins with a recap of the previous one in which Rachel (on their one-year anniversary) suggests that she and Ross should take the famous 'break'. Ross then goes to a club, from which he rings Rachel at her appartment and hears Mark (Steven Eckholdt), Rachel's boss and the source of tension between them. This precipitates an on-screen kiss between Ross and Chloe (Angela Featherstone), 'the girl from the copy place', who also happens to be in the club. The episode proper then begins, as the title indicates, the following morning, with Rachel, unaware of what has happened, intending to reconcile with Ross. She goes to his apartment and, ignorant of the fact that Chloe is still there, apologises and they make up. Soon after, however, Rachel is enlightened and following a day and night of painful and angry discussion with Ross, she concludes that their relationship has come to an end.

In the opening scene, Aniston plays the 'straight man' to Courteney Cox's Monica, providing only set-ups and no punchlines. While Cox loads a blender with fruit, she asks how the anniversary dinner went. Aniston's line, 'Well we never actually got to dinner' tees up Cox's 'Ooh nice', which gets a laugh and 'We kinda broke up instead' is the cue for Cox to start the blender without the lid, which sends the fruit flying to the ceiling. Aniston exerts little kinetic and vocal energy in this scene. In contrast to Cox's more expansive reactions, Aniston's physicality is subdued. She uses slower, stiffer movements paired with downward inflections for the majority of her lines, indicating a sombre mood. There are, however, also signs, that an optimistic viewer may wish to pick up on, that there may yet be a happy resolution. On the line, 'I realised how much I love your stupid brother', Aniston brushes some fruit off Cox's cheek with a gentle caress, exhibiting the character's capacity for tenderness. This improvised gesture (that the fruit would land on Cox's cheek could not have been predetermined) coincides with the word 'love' and seems to be motivated by it. The action is therefore apparently both psychologically motivated and determined by the material reality of the environment. These are two of the fundamental tenets of naturalist drama. (Pickering and Thompson 2013). Both Brett Mills and Cassidy and Knox argue that naturalist acting strategies can be identified in *Friends* but where Mills argues that the actors switch between comedic and naturalist playing styles (Mills 2005: 71), Cassidy and Knox argue that 'comic moments are ultimately anchored and integrated into realism/naturalism' (Cassidy and Knox 2015). As this chapter seeks to demonstrate, close attention to the details of Aniston's performance in the light of two of the main tenets of naturalist drama (that action is determined by the characters' psychology and by their environment) reveals a more complex relationship between comedic sitcom performance and naturalism that is neither entirely oppositional nor entirely compatible.

In the scene that immediately follows the upset blender, Aniston's performance is purely vocal. Chloe comes out of Ross's en-suite bathroom to his, and our, surprise. The opening credits and theme song then effectively, but temporarily, rescue Ross from the situation and the following scene begins with Ross checking his answerphone messages. Rachel has called to apologise. Again, Aniston provides the set-ups for her scene partner to get the laughs. While Schwimmer's tone has a heightened sense of remorse to the extent that he does not fully articulate the words he speaks, Aniston's tone is even and matter-of-fact. Following James Naremore (1988), Mills defines 'acting' in terms that make it synonymous with naturalism and places it in contradistinction to 'performance' (2005: 69). As Mills sees it, with 'acting' there is an attempt to efface the gap between actor and role whereas 'performance' capitalises on dual identities of actor and character, privileging the former over the latter and prioritising the demonstration of the performer's skill set over the credibility of the character. Mills sees an emphasis on excess in the comic 'mode' that better fits 'performance' as he defines it (ibid.). In her first two scenes, therefore, Aniston's style seems to fit quite compatibly with serious, naturalist acting rather than comic performance. Her movements are not excessive nor intended to highlight a sense of skilfulness. They are plausibly motivated by the character's psychological reactions and appropriate for the diegetic environment. As the episode progresses, however, Aniston's behaviour does not wholly fit the binary opposition model and raises issues with attempts to define 'acting' and 'performance', 'naturalist' and 'comedic' in contradistinction to each other.

Having heard in Rachel's answerphone message that she will drop in on Ross before going to work so that they can make up in person, Ross hurries to get Chloe out of his apartment. As he opens the door to usher Chloe out of it, however, Rachel is standing in the doorway. Aniston is primed with her hand up in a knocking position. She looks up and smiles at Schwimmer. Then, in response to his start and swift jump back, she jerks her arm, still held in the knocking position, up then down. The door is now wide open and as such, for the character, this may seem like a redundant movement chosen by the actor purely for comic effect. Yet, it is also explicable as a nervous impulse on Rachel's part, which is appropriate as a response to the peculiar reaction she has been met with. It also deftly adds to the awkwardness of the scene. With the jumpy reactions and a sense of forcedness to her movements, choices which read as credible for Rachel overcompensating for anxiety over possible rejection, Aniston contributes to the precariousness of the situation. Several times in this scene she holds her mouth in positions that suggest she is about to speak. When the door is first opened it looks as if she might say 'Hi' but then her face falls in reaction to Schwimmer's retreating movement. She mouths 'Oh' along with Schwimmer as he vocalises his response, changing her original facial composition to mirror his. Repeatedly, Aniston abandons movements that she begins to make in favour of reacting to a stimulus from Schwimmer and often paralleling his behaviour. In a very delicate way this creates a dynamic between the two characters in which Rachel's endeavours are continually thwarted by Ross. In observable, physical terms, this reflects the narrative problem. Rachel has finally been given the opportunity to work in the fashion industry and achieve the job satisfaction that she has been searching for over the last two-and-a-half seasons (having been an unsuccessful and unhappy waitress at the coffee house). Ross's insecurity about her having a male boss has blighted this achievement and put a strain on their relationship, causing Rachel to propose the 'break'. Here, on asking whether she can be his girlfriend again, Aniston searches in Schwimmer's eyes, which works to increase the sense of uncertainty about their future together. The lack of physical ease between the actors at this point suggests the characters' discomfort with each other and sows a seed of doubt over their compatibility. Aniston makes a substantial contribution to the mood of the scene through bodily disclosures of tension that are highly responsive to her surroundings. The accumulation of these details lays the groundwork for the break-up.

As Monica and Phoebe (Lisa Kudrow) get excited about a new waxing product that has just arrived by mail order, Ross is next door at Chandler (Matthew Perry) and Joey's (Matt LeBlanc) apartment trying to figure out how to tell Rachel about Chloe. Chandler and Joey convince him not to and Chandler lays out 'the trail' from Chloe to the unsuspecting Rachel. This sends Ross on a search for the people who are possible links in a gossip chain that will end in Rachel finding out. Eventually Ross goes to Gunther (James Michael Tyler), who works at Central Perk, the coffee house. Unbeknownst to the characters, but well known to the audience, Gunther has had a long-time crush on Rachel and has a clear motive to tell her about Ross's infidelity. When Ross asks him if he has said anything to Rachel, Gunther replies 'Was I not supposed to?' in a tone of feigned innocence that is apparently lost on Ross. This is punctuated by laughter from the audience but the mood swiftly changes as Schwimmer turns around and the camera pans right to reveal that Rachel is sitting by the window. Aniston's eyes are fixed on Schwimmer. She holds onto a bag with both hands as if not willing to let go, which contrasts with her later open-palmed hand movements that attempt to bat Schwimmer away from her. Here she does not move a muscle. The stillness and lack of comfort in her posture convey a sense of shock, creating the impression that Rachel has been frozen in this position for a while, stunned by the news, unable to function, or unwilling to operate in the post-revelation world. This image foreshadows her words to Ross at the end of the episode that she had thought of him as someone 'who would never, ever hurt me. Ever.' Her rigid position also has particular resonance after the previous scene in which Aniston's attention to Schwimmer is so steadfast and the impact of his behaviour on her is such that his body language determines hers.

There is a cut to Monica's bedroom. Hearing Monica and Phoebe's waxing-induced screams, Joey and Chandler come running into Monica's bedroom brandishing a saucepan and a kettle as make-shift weapons of defence. Then, when they hear Ross pleading with Rachel to talk to him and Rachel responding that she can't even look at him right now, Chandler, knowing why they are fighting, shuts the bedroom door and, in doing so, confines the four friends to the position of secret audience for the duration of the break-up.

Aniston ineffectively tries to shut the apartment door on Schwimmer and, at the infuriating futility of this, voices a pursed-lipped exhalation. Then Aniston strides away from Schwimmer, swatting her hand backwards and then downwards in two forceful motions that physically reinforce her verbal command for him to move away from her. Compared to the earlier gentle touch on Cox's cheek, there is a hardness to Aniston's movements that marks a different side to Rachel's temperament. This retreating physicality is quickly replaced by that of confrontation. In response to Ross's rationalisation that he 'made a mistake', she faces Schwimmer squarely and throws the words back to him. 'A mistake?' She pauses. 'What were you trying to put it in?' She pauses again. 'Her purse?' This is delivered in a very high, loud, screeching voice with two exaggerated inflections marking the punctuation. Aniston uses the top end of her vocal register, tossing the words up in a deliberately uncontrolled manner to express Rachel's exasperation. Aniston uses highly emphatic mouth movements to enunciate the words, which results in a large expanse of her teeth being visible, displaying Rachel's anger through body language. Aniston puts the emphasis on 'trying', elongating the word, which, between the staccato monosyllabic words, establishes a rhythm. As the rhythm is left incomplete by the end of the grammatical sentence, it signals that she will provide the answer to her own question. Aniston does not upwardly inflect 'put it in?' because this would vocally pass the ball to Schwimmer and it is not the end of Aniston's performance segment. Instead she holds off from raising the pitch of her voice, prolonging the anticipation for the punchline until 'her purse', when she now modulates up and completes the rhythmic pattern, thus giving the aural cue to the audience that it is time for their response.

Here, Aniston's movements may perhaps be characterisable as 'excessive display' (Mills 2005: 69) and her vocal choices may draw more attention to her control of her delivery, yet neither are incompatible with Rachel's psychological motivation. For a more nuanced understanding of the relationship between comedic performance and naturalism, the theatrical conventions of both require further elaboration, for, as Cassidy and Knox (2015) indicate, excess need not jeopardise a naturalist aesthetic. Rather, I would argue that the feature of conventional live comedy performance that is most directly in tension with a naturalist aesthetic is the covert, yet visible and audible, acknowledgement of the audience. It is the reconciliation of the appeal to the audience and accommodation of their aural presence, which live comedy requires, with the disavowal of the audience, which naturalism dictates, that Aniston manages so adeptly.

On Rachel's direct verbal articulation of the fact that Ross has had sex with another woman (which is necessary exposition for Phoebe and Monica) Aniston lurches towards Schwimmer. This is the first move closer to him after the moment of anagnorisis and begins a pattern of physical advance and retreat, which visually demonstrates and inverts the 'will they? won't they?' dynamic. The question is no longer the familiar sitcom romance storyline, 'will they eventually get together?'; it is now 'will they suddenly break up?'. Moreover, Aniston's movement towards Schwimmer is also a movement towards the audience, which is highly significant in terms of the relationship between comedic intention and a naturalist aesthetic. At the same time as jutting her head and making a transfer of weight in Schwimmer's direction, Aniston also keeps her downstage shoulder back, thus positioning herself on a diagonal axis, 'cheating out' to the audience. Throughout this scene Aniston uses the set to protract her body along the diagonal. Asking Ross to leave, she opens the apartment door with her downstage hand and stretches across to the breakfast bar with the other, presenting the front of her body at a 45-degree angle to the studio audience. Then, when Ross wants to talk about what has happened, she slams the door and screeches 'How was she?' as she positions herself upstage of the dining table, placing one hand on the table and the other on her hip, in another posture that opens up her body to the viewer. Her position on the soundstage is often, therefore, not solely determined by Rachel's psychological state or the diegetic environment but also geared towards accommodating the studio audience. Aniston places her body so that it is midway between the audience and Schwimmer as an invitation to the audience to play an active, albeit limited, role by contributing laughter to the aural aesthetic of her performance. This is in direct tension with naturalism's fourth wall that separates and ignores the meta-theatrical realm.

An acknowledgement of the non-diegetic human presence is instrinsic to the aesthetic of sitcom performance recorded in front of a live studio audience. It is Aniston's skill for managing the threat that extra-diegetic activity poses to the credibility of the diegetic world that makes her particularly adept at this performance style. While positioned towards the audience, Aniston roots herself with the décor, maintaining a tangible connection to the character's environment, and holding eye contact with Schwimmer. This suggests strong psychological engagement with the action within the pro-filmic space. As such, Aniston's body language both effectively conveys a sense of Rachel's reality and registers the audience's presence. Open postures enable Aniston to comfortably combine the adherence to both the frontal blocking of conventional live comedy and the psychological motivation of naturalism.

The more problematic issue for the compatibility of live comedy and naturalism is that frontal blocking tests the boundaries of the naturalist hermetic enclosure of the diegetic realm. The question at the heart of this type of sitcom performance is how far can the actors risk playing out to the audience before jeopardising the credibility of the fictional world? Non-comedic acting plays a vital role here. The more robust the actors' manifestation of multifaceted and emotionally rounded characters, the more audaciously they can ask for the laugh. The more the actors shore up the characters' believability, the greater the definition and dimensionality they give to the diegetic realm and this affords them greater liberty to appeal to a presence beyond the fourth wall before it is compromised.

In this episode Aniston physicalises a spectrum of emotional pain from the initial inertia in the coffee house, through frustation, up to rage and ending in exhaustion. The greater part of the emotional arc that Aniston performs takes place solely in Monica and Rachel's open plan kitchen and living room. Making full use of the set, props, her co-star and her own instrument, Aniston builds from a slow-burning fury to an energetic expression of rage to a state of depletion.

In response to Ross's choice of the word 'Different' as his answer to 'How was she?', Aniston precisely enunciates '*Good* different?'. Adding a little pause between the words so that both 'd's are heard, she spits out the 'd' of 'good' and the 't' of 'different' as she narrows her eyes. Her lack of modulation within the words themselves gives them a serious, almost dangerous tenor, but she places 'different' on a higher note, so that there is an upward inflection, which shoots the question to Schwimmer. Aniston has not changed her physical position, but through lowering her voice she creates a narrower performance area, concentrating the dramatic tension within the pro-filmic space. She takes a beat to punctuate his response 'Nobody likes change', which is the punchline, so that Schwimmer can get the laugh before she grabs the newspaper on the table to start hitting him. Although this conventional comedic business (characters hitting each other with props is familiar from Punch and Judy Shows) receives one of Aniston's biggest laughs of the episode, there is a forcefulness to her frenzied strikes that imbues the action with a sense of genuine rather than comedic rage.

As Rachel processes the information that Chloe was still in Ross's apartment when she went there that morning, Aniston adopts an increasingly extreme tone. She begins by vocally arresting Schwimmer with three drawn-out 'Woah's. Then, in a measured 180-degree rotation that spans one end of the studio audience to the other, Aniston turns around to confront Schwimmer, holding her hand in a gun-like position (see Figure 11.1), verbally and corporeally 'holding him up'. But once the information has sunk in, she grabs her stomach. It is Rachel who has been wounded (see Figure 11.2). Aniston then cups her hands to her mouth as she asks, 'She was there?', seemingly unable to release the words yet. Then she repeats

Figure 11.1   Aniston 'holds up' Schwimmer . . . (*Friends*, NBC, 1994–2004)

Figure 11.2   . . . and then demonstrates it is Rachel who is wounded (*Friends*, NBC, 1994–2004)

them with her hands over her heart adding 'still'. Holding her heart and her stomach are recurrent gestures Aniston uses for Rachel. They form part of what Philip Drake refers to as an 'idiolect', 'a repertoire of performance signs [...] associated with a particular actor' (2006: 87). I believe that the visceral reaction these gestures suggest are key to why and how we feel when we watch Aniston. Through physical poses such as these, she visibly signals internal pain. When she hits fever pitch with a third iteration of the question she lurches her head forward and Schwimmer offers her back the newspaper to hit him again. Aniston grabs at the offering with both hands and a complete transfer of weight towards Schwimmer as she mounts the sofa, squealing 'Just get away' with such a shrill timbre that the words are almost unintelligible. The second time round, hitting Ross with the rolled-up paper is not an impulsive action for Rachel but an opportunity for her to demonstrate her anger. Rachel is cued by Ross to repeat her action just as Aniston is cued by Schwimmer. The excessiveness of the movements and vocal delivery therefore coincides with the character being prompted to perform. Within seconds of screen time Aniston makes rapid, dynamic shifts in the energy of the scene, building tension, demonstrating hurt and then enabling a sense of release through a move into an aesthetic reminiscent of classical farce as she precariously tears around the set and over the décor.

On hearing the news that Ross tried to prevent Gunther from telling her about Chloe, Aniston releases a held breath, relaxes her posture and softens her face. There are two main audience sounds that can be distinguished here. A polyphonic chorus of high-pitched intakes of breath is followed by a wave of low 'woahs'. There may of course be a false ontological link between the audience laughter as it appears in the episode and Aniston's performance as it was given in the pro-filmic space. While the studio audience reaction is highly influential in terms of timing for the actors and gauging the comedic value of lines for the writers, in terms of performance analysis it is not appropriate to credit the actor with achieving the aural appreciation that we hear on the soundtrack, as the laugh track is manipulated in post-production to reinforce the signal for comedic intent and not necessarily representative of the studio audience's response to the actors' performance. In *Friends: The One that Goes Behind the Scenes* (Discovery Channel, 1999), a documentary on the making of an episode, the series editor explains that if a live response is too long for television, he replaces the studio laughter with a carefully selected 'LAF' track which ranges from 'short chuckle' to 'long w/applause'.[2] Whether the gasps and ominous sounds of disapproval are genuine or edited in, Aniston's delivery of the line 'Oh that is so sweet' is at odds with them. The audience seems to judge Ross's attempt to cover up his actions negatively, whereas Rachel seems to be deeply moved by it.

Then there is a change in Aniston's body language and tone of voice that indicates a move from an appearance of a simple one-to-one relationship between her mind and her body to a physicality that signals that she intends the audience (and Rachel intends Ross) to recognise that she does not mean what she says. The sincere register that Aniston begins in gradually becomes increasingly sarcastic. She adds greater strain to her voice; exaggerates her mirthless smile; takes a beat before speaking, suggesting deliberation over her word choice. Aniston teasingly becomes slightly overemphatic on the line 'Oh that is so sweet', which starts to bring Rachel's sincerity into question. Then after the line, she strikes her hands-on-heart pose, which is apt as its status as a stock gesture adds to the sense that Rachel is showing only assumed sincerity. But there is a deliberate ambiguity here, which works as a stepping stone to the more transparent sarcasm to come. The exaggerated rising and falling cadence of 'I think I'm falling in love with you all over again' conveys exasperation, further emphasised by the matching descending hand gesture. The reveal that Rachel is not in fact being sincere is a valuable performance tool, enabling Aniston to reconcile Rachel's hurt with her opportunity as an actor to induce laughter from the audience. Aniston first employs her naturalist acting skills to appear earnest, then fully exploits for comic impact the rapid switch in tone that comes with Rachel's revelation of inauthenticity. Aniston's demonstration that Rachel is being what Erving Goffman terms a 'cynical performer' (1959: 29), an individual who is not taken in by their own social performance, facilitates a smooth transition from exaggerated movements that suggest calculation for maximum effect on an audience to a sense of unfiltered expression of feelings. What is important to emphasise is that the distance that Aniston creates between performance layers is not between herself as a performer and Rachel as a character in the way that Alex Clayton describes the 'comic twinkle' (Clayton 2012: 51). Aniston does not highlight a discrepancy between actor and role, but through the act of exposing Rachel's self-conscious facade, she deconstructs a superficial performative level involving a greater degree of self-awareness yet remains securely within the bounds of a single identity that is consonant with the diegesis, i.e. the person who we understand as Rachel. As sarcasm relies on the ability to use body language and tone of voice as a metacue to contradict linguistic communication, Aniston can showcase her vocal performance skills without putting strain on the credibility of the character or the situation. With this, Aniston can legitimise her performativity and the comedy can result from a disclosure of inauthenticity on the part of the character rather than the actor.

Susan Berridge (2014), referencing Karen Lury, suggests the notion of a blurred distinction between actor and character when it comes to television performers is particularly applicable to Aniston. I would argue that here, as Aniston reconciles the motivations of character and actor by uniting them in one action, she works to prevent a sense of incongruity between 'Jennifer Aniston' and 'Rachel' from posing an imminent threat to the viewer's willing suspension of disbelief. More problematic for the maintenance of a naturalist aesthetic than the notion of dual actor-character identities is the tacit and subtle but understood acknowledgement of the studio audience that comedic playing incorporates. The timing of 'I think I'm falling in love with you all over again' is determined by the intention to elicit laughter from the studio audience. Aniston takes a beat before and after the line to isolate the comic material and signal that that is what it is. Her trademark hand-on-heart pose at the end of the line then enables Aniston to wait for the duration of the laughter in a position that seems natural and comfortable so that it doesn't look too obviously like that is what she is doing. Aniston is employing the sarcastic tone as a means to induce laughter from the audience; but there is more to Rachel's use of it. The physical advance and retreat motif that Aniston employs is mirrored here at the psychological level. It feels appropriate to interpret Rachel's use of sarcasm as a distancing device to shield herself from Ross as throughout the episode she oscillates between emotional openness and guardedness. The complexity of the psychological motivation adds depth to the character that balances the actor's accommodation of a presence outside the diegetic environment. By heightening one element of a naturalist performance aesthetic (psychological motivation) Aniston compensates for the fragile status of another (the fourth wall).

After a transition that signals the passing of time, Aniston conveys the weariness of having fought all day through limp wrists and a lack of vocal energy. On her hypothetical suggestion that, say, she had slept with Mark, which has been a long-standing fear of Ross's about which he has been unable to be reassured, Aniston wearily rotates her head, physicalising her waning stamina to take part in the same fight the characters have engaged in over several episodes. This is the last time she walks towards him. In the final scene of the break-up, Aniston plays solely for empathy and not for laughs. She encourages us to believe that there is no longer a gap between her internal life and her physical expression. As we are denied the release of tension through laughter, we are positioned to feel Rachel's pain more acutely. Aniston has to show that Rachel is in the process of psychologically extricating herself from the relationship. She has the same tools at her disposal as in the scenes in which Rachel receives and digests the information that Ross tried to prevent Gunther from telling her and that Chloe was still in the apartment at the time that she was there in the morning: the dialogue; her body; her co-star; the pro-filmic space; the studio audience; and the cameras. But she uses them very differently. Aniston no longer invites audience participation. She speaks in a quieter tone, restricting her energy totally to the space within the set. She lets out a quiet whimper, and then an intake of breath mutes a cry. 'No' is said very softly and gently so that it has an apologetic, almost pleading quality to it. Then she pauses. Her mouth is shaking. She sniffs to pull herself together before she says, 'I can't'. The almost whispered words are swiftly followed by a low and even but rapid delivery of 'you're a totally different person to me now'. She then returns to a whisper for 'I think you should go'. It is clearly very difficult for Rachel to say, but she is resolute. Aniston bats Schwimmer away gently this time, in an indirect and light movement that suggests resignation rather than attack. She brings back the softness she exhibited in the first scene of the episode. She returns to sotto voce for a measured and enunciated 'I used to think of you as someone who would never ever hurt me'. In response to the question 'This can't be it?' Aniston breathes out and replies with a question of her own: 'Then how come it is?' Aniston makes eye contact with Schwimmer and ends on an upward inflection. It is not delivered as a rhetorical question. Rachel *is* asking Ross a question, but one that is answered through a lack of action. His silence takes on the meaning of a sombre acquiescence to the end of the union.

Throughout the series Aniston often affects crying, signalling to the audience through a comic twinkle that Rachel is upset but that they should be finding it humorous. For example, in season three, episode ten, when Rachel cannot get her job at the coffee house back as she has been replaced by a new, much more proficient waitress who is able to make paper napkins into swans, Aniston only gives a vocal performance of tears, which is discernibly at odds with her relaxed physicality and as such clearly detectable as simulated for effect. Aniston says the word 'swans' unintelligibly high as she covers her eyes with her hands, which simultaneously suggests tears and conceals the fact that she does not actually have tears in her eyes. Here, however, Aniston's quivering lip appears not as a performative signal but seems to be a physical consequence of psychological distress. The apparent causal link between Aniston's physicality and her inner state fosters empathy with her feelings rather than appreciation of her command of her body. The sniffing that can be heard in Aniston's voice suggests that Rachel is trying to gain control over painful and uncontainable emotions. Rather than giving the impression of trying to cry, which would suggest crocodile tears as with the above-noted example, Aniston's performance secures the impression that Rachel is trying *not* to cry. The appearance of prohibiting emotions from reaching the surface suggests their depth. Eschewing physicality and uses of voice that are more overtly directed at an audience, Aniston now removes the element of comedic performance that was in playful tension with the naturalist hermetic concealment of the diegetic realm. Whereas earlier in the episode Aniston employed her whole body in forward motions directed towards Schwimmer, using open palms and bared teeth to indicate that Rachel was prepared to engage with Ross, here her body has returned to the state of inertia seen in the coffee house. Limiting her physical expenditure, Aniston makes a clear contrast between Rachel's transititional emotional state in the middle of the episode and her now firmly changed feelings towards Ross. Postural changes are only made in a retreating direction and her gestures block her body rather than opening it out. Aniston regulates her breathing to show that the decision to end the relationship is not one that Rachel is making in the heat of the moment, which could be easily reversed by the next episode. This is a sad and sober realisation that precipitates change.

Attending closely to Aniston's use of physical and vocal expression, her placement within the set and her use of props, I have examined the relationship between her appeals to the studio audience for the laugh and two elements of a naturalist aesthetic: psychological motivation and the hermetic concealment of the diegetic realm. Through detailed analysis of the tonal shifts in Aniston's work in this episode, I have offered a reading of sitcom performance that appreciates a more complex interplay between an intention to induce laughter and the task of maintaining the reality of the diegetic world than is implied by either a clear-cut dichotomy of 'comic performance' and 'naturalist acting' or the notion of their total compatibility.

Notes

1 On the duality of actor and role in comedic performance see Clayton (2012), Mills (2005) and Naremore (1988).
2 The documentary is available on YouTube: www.youtube.com/watch?v=J4qSQxE4_Y4 (accessed 26 April 2019).

12

Soft Upper Lip: Coach's Facial Expressions in *Friday Night Lights*

Timotheus Vermeulen

This chapter takes a closer look at Kyle Chandler's performance as coach Eric Taylor in the NBC drama series *Friday Night Lights* (NBC, 2006–11), in particular the movement of his mouth in relation to the expression of his face. Drawing on close textual analysis, film theory and Gilles Deleuze and Felix Guattari's conceptualisation of faces as landscapes, I argue that Chandler tends to employ three interrelated registers, or frames, of expression within which most affects are articulated; three linked landscapes, if you will, in which the action takes place. These are: the 'scroll plain', in which parted lips hang still in an otherwise still face; the 'floodplain', in which the bottom lip is bitten, drawing the whole face towards the mouth; and the 'outwash', in which the lips are pressed stiffly together, pulling the whole face towards the chin. Of course, Chandler employs other, more dramatic registers as well, but these are the most frequent. Examining instances of each of these variations, the argument is not that these frames prescribe, caricaturally, the whole gamut of Taylor's affective range, let alone all human emotion, with each frame matching a specific sentiment. Rather, I take them to circumscribe specific spaces in which much of the character's emotion can come to the fore, reliefs, you might say, which allow for some emotions more than others, just as certain landscapes – plains and polders, river beds and beaches, valleys and mountain ranges, woodlands and deserts – encourage different activities. As such, the chapter proposes – or in any case probes the possibilities and limitations of – a new model for studying television performance, specifically facial expression. It offers a preliminary (and by no means exhaustive) categorisation of facial

225

expressions as distinct spaces, with discrete histories, with specific narrative conventions and aesthetic modi operandi; landscapes whose meaning is neither completely fixed nor entirely flexible but endlessly established and re-established – or, in the terms of Deleuze and Guattari, de- and reterritorialised – along certain axes of signification, such as narrative and character development, both across seasons and/or episodes and within individual episodes, tone, and visual context, in other words cinematography and mise-en-scène.

A Plain Face

To say that *Friday Night Lights*, a drama series documenting five or so years in the life of a Texas high school football coach, his family and his students, delivers on its generic contract – drama – would be an understatement. The series' dramatic range is both expansive and intensive, covering a widening variety of emotions in ever quicker succession. It also encourages theatrical performances, from the adults as much as, as may be expected, the teenagers. Indeed, there are so many last-minute plot twists, so many grand gestures of joy, so many voices quivering with anger, and so many melodramatic tears, that the *New York Times* critic James Poniewozik (2016) at one point described the show as a 'cry-time drama'. It may seem surprising, in this respect, that the performative register of the show's central character, high school football coach Taylor, is rather subdued. It could be suggested, however, that it is because the series is so melodramatic that its protagonist is so restrained, the pivot around which everyone and everything else orbits, an anchor keeping the fictional universe from floating away. No matter how stressful the occasion – financial worries, familial trouble, a student in need, an important game played poorly – Taylor's body language tends to be inhibited, his gestures minimal and his facial expressions minute. He rarely screams, while I cannot recall having ever seen him laugh out loud. There is an apt expression in Dutch, my native language, which says that you shouldn't ever show the back end of your tongue, which means being reticent, keeping your thoughts and feelings to yourself. Taylor literally never shows the back end of his tongue. This is not to say that he strikes a composed figure, or is that most mythical of characters, the suave 'strong and silent' type. There is nothing smooth about him, no sense of serenity. Indeed, I would suggest that, on the contrary, there is an ever so slight edge to him that he cannot seem to shake, a subtle tension, a faint testiness, that inhabits his every move and manner, which anticipates his moments of agony as well as circumscribes those of pleasure, as if his body constantly fears the weight, or, yes, the levity, of the world, tipping; as if every expression is met by pinpricks he attempts to avoid. People hide the back end of their tongue, after all, not because they are secretive, but because they don't know whether to trust those around them.

Few moments in the series illustrate Taylor's reticence – and his environment's expressiveness – more manifestly than the closing minutes of the final episode of the fifth (and final) season. With Taylor's family considering moving home and his team playing the state final, there is a lot going on – for everyone. Indeed, the intensity of the moment leaves its marks on all the characters around him: his wife Tami (Connie Britton) alternately crying and laughing, her arms closing and opening up, the features of her face contracting vertically and broadening horizontally; his fellow coaches running up and down the line, gesticulating heavily as they scream at the top of their lungs, their foreheads sweaty, eyes red; the players slapping each other's shoulders and pumping their fists, cheeks blowing and mouths wide open. Taylor himself, however, barely moves a muscle. His bodily composure and facial expression the moment he informs Tami that he will join her in Philadelphia (a decision he has doubted for episodes, the delay of which has caused considerable friction with his wife) do not differ much from how he looks while telling his star quarterback Vince (Michael B. Jordan) how proud he is of him, or watching his team score a touchdown in the State final, or seeing the other team win a point, or observing an uninspired training exercise months later. In each of these instances of varying emotional range and intensity, he stands or sits still, ever so lightly bending forward, leaning on the front foot; his forehead is unwrinkled, his eyes stare at the other's intently, lips are pressed together, the corners of the mouth turning slightly downwards as the chin points marginally forward. Of course, there are subtle differences: before sharing his decision with Tami, he momentarily holds back, looking downwards, as if to find not just the words but also his composure. While telling her, his eyes focus and though his tone remains level his articulation picks up speed, both betraying a nervousness and signalling significance. Right after, his eyes turn left and right and up and down in an apparent attempt to look for signs as to what his wife is thinking. His confession to Vince is shared with a quick but pointed raise of the eyebrows. After a difficult few minutes at the game, the camera catches him as he briefly bites his lip; at another point during the game, he is seen silently grinding his teeth. But the sense that prevails, across all of this, is the extent to which Taylor is, especially in relation to everyone else, restrained in his expressions. Indeed, as if to make sure the point is driven home to the viewers, the show at one stage cuts from close-up to close-up to close-up, face to face to face, from Taylor's to his wife's to that of his daughter and her boyfriend to his friend Buddy to Vince's father to that of his mother and so on, as they all watch the quarterback's decisive pass fly through the air, with Taylor's face obviously the odd one out, the only one not to overtly raise his eyebrows or gaze in awe or open his mouth or grimace or laugh (Figure 12.1).

What I am getting at here, I suppose, is that Taylor's face is plain, that he has a plain face. I mean that his face communicates little of his emotional state, or much less, in any case, than the faces of those around him do. But also that it resembles a plain: a relatively flat area of land, often but not always surrounded by – and, as certainly in Taylor's case, offset

Figure 12.1 Coach Taylor stiffens his jaw in anticipation (*Friday Night Lights*, NBC, 2006–11)

against – more and less elevated patches of land such as mountain ranges or valleys. Geologists distinguish between three types of plain: the structural plain, which is a fundamentally depressed area; erosional plain, which is an area levelled by rivers cutting through it and wind sweeping across it; and the depositional plain, in which the flatness is the result of sedimentation. Taylor's face, I would be inclined to say, resembles this third plain, his skin thickened over time, the consequence, as the show suggests throughout, of a lifetime of temporary jobs and short-term leases, of fathering a girl that outsmarts him and coaching volatile teenage boys, of being a local role model and a community punching bag. Indeed, if you compare Chandler's performance in the pilot episode and the final instalment, it becomes apparent not just how much weight Taylor has gained, but how much of his initial expressiveness he has lost – which, to be sure, wasn't too theatrical to begin with, but, in its occasional intimations of surprise or joy (the mouth widening, the eyes narrowing) or, especially, irritation, anger (the mouth turned into a grimace or a scream), is a different landscape altogether from the one left by the show's end. I feel I should say here that though I make the comparison between face, or facial expression, and landscape with some apprehension, since it runs the risk of reducing context-dependent complexities to a general type, I do intend it as more than a gimmick. To suggest a face can be studied like a landscape has implications for the types of question we might think to ask of it – which is, to be sure, what I am interested in here. It allows us, for instance, to contemplate its formation, to reflect on expressions as the consequence not just of immediate events but of protracted processes, some of which we may have witnessed, others which we might infer. It further offers insight as to what could and what could not feasibly happen – which movements would deepen or widen the illusion and which may wilfully or accidentally break us free from it. I believe it helps us understand Taylor's expressive register, the movements of his mouth or the lines around his eyes, the tension in his jaw or the looseness of his cheeks, if we can decide whether his face is a structural plain or an erosional plain, an erosional plain or a depositional plain, to read his face as more than a response to the demand of a particular plotline, to read it, instead, or also, as the consequence of narrative developments more generally, or indeed, of living – of interacting, of thinking, of feeling, going through the motions – in this fictional universe.

The Moving Image as Still Life

In describing Taylor's face as a landscape, I take my inspiration from Deleuze and Guattari's discussion of faces as landscapes and landscapes as faces in *A Thousand Plateaus* ([1980] 2016). Though I draw on the comparison both far more literally and slightly more metaphorically than they do, taking literally a metaphor they introduce to explain a concept – deterritorialisation – I do want to say a few words about their argument, especially in relationship to Deleuze's later study of faces in cinema ([1986] 2005). The face, Deleuze and Guattari argue, is not a natural feature of the human body. In fact, they suggest, the face exists exclusively in the simultaneous abnegation and replacement of the human body. As they put it:

> the face is produced only when the head ceases to be part of the body, when it ceases to be coded by the body, when it ceases to have a multidimensional, polyvocal corporeal code – when the body, head included, has been decoded and has to be *overcoded*.
>
> ([1980] 2016: 199; original emphasis)

What they mean, in other words, is that what we tend to call the face is the beheading of the body as organism, as organic matter, the human as animal, and the head's reattachment in another context, one that, as the repeated inclusion of the verb 'to code' suggests, operates along the lines of signification and subjectification – the face as communicative surface. The narrowing of the eyes and the widening of the mouth, the snort and the cocked jaw are figures of speech, indices not of the presence of your own body as much as evidence of that of someone else's. In this respect, Deleuze and Guattari note, the face is to the body what the landscape is to nature: a picture of nature cut out from the natural sequence which in another frame comes to connote it, that is to say the decontextualisation of a part from the whole, or one wave from the drifts, one tone from a composition, where – in the novel configuration – the part displaces that whole in the process (2016: 201). The label they attach to this process, of course, is de- and reterritorialisation.

In *Cinema 1: The Movement Image*, Deleuze further develops this understanding of the face as deterritorialisation of the body as 'movement of extension' and its reterritorialisation as 'movement of expression' ([1986] 2005: 90). I guess you could say that if in *A Thousand Plateaus* the face is cut from the body, here he severs it from the social, focusing exclusively on the face's formal features in and of themselves ([1986] 2005: 105). It is especially the close-up that draws his attention. The 'close-up', he writes, 'extracts the face from all spatio-temporal coordinates' ([1986] 2005: 111), 'opens up a dimension of another order favourable to these compositions of affect' ([1986] 2005: 104). There are, Deleuze suggests, two questions we can put to a face – or, speaking about cinema, two questions that the close-up can, and as a rule does, answer for us: what is it thinking about? And: what is it feeling? ([1986] 2005: 91) In the first scenario, Deleuze writes, we tend to talk to the face's outline, to the face as calm unity, where each of its disparate elements function harmoniously as one. To be sure, formal harmony here does not mean that the content is harmonious, not necessarily in any case. Indeed, Deleuze cites wonder and surprise but also melancholy and solemnity, a 'funereal glacier' ([1986] 2005: 92). He calls this perspective the 'reflexive face' since it *reflects* a pre-existing thought, or, in cinema, theme, or narrative, or tone; a state, a '*quality*'. The second question, says Deleuze, on the contrary, addresses a face's individual features as they split apart, restlessly, chaotically, now the lip quivering, now the eyelid trembling, then the veins pulsating. Here we don't see a face, in fact, but movements, paroxysms. This perspective is labelled the 'intensive face' because the face is, in effect, a series of intensities – the lip that starts to quiver, the eyelid that begins to tremble, etc. If the reflexive face expresses a state, a 'quality', the intensive face expresses the transition, the transformation from one such quality to another, or a series of others. In this respect, for Deleuze, faces in cinema, faces in close-up, communicate either stasis or movement, illustration or creation, being or becoming – or, since these are categories of one and the same face, a combination of both in differing hierarchies.

It is interesting, I think, that though the discussion of the face in *Cinema 1* is in many a sense the continuation of the conceptualisation in *A Thousand Plateaus*, Deleuze doesn't pursue the comparison with landscape any further – which is not to say that he doesn't occasionally describe shots in terms of landscape (indeed, the metaphor of the 'funereal glacier' is one of a few), but rather that he doesn't generalise it. I assume the reason is that Deleuze theorises the close-up not just as the representation of the face, but as a face itself: the close-up makes whatever it concentrates on into an interaction between a still unity and isolated movements, abstracted from time and space, a world onto itself. Strictly speaking, another type of shot does not turn its content into a face the same way. Certainly not, you'd imagine, the shot that may well be considered the close-up's opposite: the wide shot and the long shot conventionally employed to depict, to call into existence, landscapes.

I don't want to argue here that the close-up of the face and the wide or long shot of the landscape are the same, since the differences between them are obvious. As Mary Ann Doane has noted (2003: 108), the close-up of the face monumentalises the minute, whereas the wide or long shot of a landscape shrinks the gigantic; the close-up draws close whereas the wide or long shot distances; and while the former turns the visual into the haptic, the latter represents texture as surface. There are, however, overlaps and resonances. Both the close-up of the face and the wide or long shot of a landscape concentrate on the tension between the immobile and movement, immobilising (the face as paralysis of the body, the landscape as still life of nature) so as to make manifest micro movements (the trembling eyelid, leaves falling); both transform part into a whole (the face represents the person, the landscape comes to equate nature); and both abstract this whole from time and space, turn their content into a spectacle suspended from and suspending narrative progress, an articulation of, yes, itself, affect – as either confirmation of what preceded or sensation as to what may ensue. Indeed, reading Jean Luc Nancy's influential discussion of landscapes ([2003] 2005), it is striking just how similar the terms of description are to those used by Deleuze in talking about close-ups. Landscape, says Nancy, is 'an immobilization in which [...] movement is grasped' ([2003] 2005: 61) – the movement of seasons or the time of day, for instance. It further 'contains no presence; it is itself the entire presence' ([2003] 2005: 58). In televisual (or cinematic) terms: landscape is not a setting. In an image of a landscape, all that 'would possess any authority or capacity for sense', that is, story, is momentarily excluded, abnegated. 'This means that the landscape can be neither theological nor political, neither economic nor moral' ([2003] 2005: 59). The landscape is 'not a view that opens onto some perspective. It is on the contrary, a perspective that comes to us, that rises from the picture and in the picture' ([2003] 2005: 59). Landscape 'opens up onto itself', its creation taking place 'ex nihilio' ([2003] 2005: 60). Finally, Nancy suggests, a 'landscape is always the suspension of passage' ([2003] 2005: 61), a moment taken from time, a slice cut from space, the opening, you might say, of another dimension. In other words, if the close-up of the face investigates the composition of affect, so does the landscape, showing, as Martin Lefebvre, another scholar of landscape, has put it, just how many 'possibilities' for emotional expression there are (2006: xii). Tellingly, both Deleuze and Lefebvre draw their inspiration from the writings and work of the same film-maker: Sergei Eisenstein. To reiterate my point: while discussing faces in terms of landscape limits the possibilities of engagement on one level, since there are so many aspects of faciality that looking at landscapes will not elaborate, it also offers fertile grounds for exploration on another. In this sense, to treat Taylor's face as a plain, a moment in a spatio-historical process of nature extracted from that process, allows us to consider not just the quality of the reflective moment, of the outline, of the face in formal harmony, but also the possibility for and power of micro movements. After all, leaves falling from a tree in a lush, green plain will have a considerably different impact, visually, than those same leaves falling in a barren, sandcoloured plain; similarly, it matters whether a river runs its regular course or floods its banks. In the remainder of this chapter, I look at Taylor's expressive register through the lens of three overlapping but distinct landscapes: the scroll plain, the floodplain, and the outwash.

Taylorscapes

For a period in the 2000s, there was a tendency among films and series, especially fantasy and science fiction features, to introduce their fictional universes from a mechanical bird's eye view, or more accurately still, the gaze of a cyborg swallow, sweeping across continents and oceans hazy from the pace, dipping in and out of landscapes emerging into view like islands from the mist. I cannot remember where I first saw it, but the last instance was in the opening minutes of the children's animation *Sing* (dir. Garth Jennings, 2016) – which goes to show just how far this tendency has travelled – where the camera flies across the fictional world at such speed as to render it invisible, only to immerse itself into the life of a distinct site for a few seconds before taking off again. A similar technique is used in *Friday Night Lights* with respect to the faces of characters, though, since we already find ourselves in place, on earth, obviously at a decelerated tempo and limited mobility. The camera often scans the scenery for faces to depict – indeed, does so increasingly as the show progresses. Moving ever so slightly up and down, left and right, while staying put at a single depth of field, everything in view turns blurry, fuzzy, except for the moments where it happens to capture a face, matching its predetermined focus with its intended subject matter, an anticipated accident. Like the landscapes in those fantasy and sci-fi films from the 2000s, or indeed, like the sites in *Sing*, these faces, too, emerge into view like islands from the mist – that is to say, the faces and their features are introduced to the viewer as points of reference and connection, worlds onto themselves, where their gaze can come to rest: landscapes, or, with respect to our object of study here, Taylorscapes.

Depending on the narrative state as well as the character's development, Taylor's face tends to emerge into view in either one of three configurations: a still face with slightly parted lips; a face contracted by a biting of the lip; or a face drawn downwards by stiffly pressed lips. As I noted in the introduction, there are other compositions, certainly, some of which I will mention in passing below, but they are either far less frequent or secondary, that is to say, deviations from those faces that we are shown initially. The first of the three, which one may call the scroll plain, a plain through which a river runs, or babbles, meanderingly, really, at a low gradient, is the composition that features most often, especially in the early seasons but in later seasons as well. In most cases, this face is characterised by a smooth forehead, straight eyebrows, forward-looking eyes, relaxed cheeks met midway by a marginally tightened jaw, slightly parted lips and a chin pointing downwards, elongating the face. On occasion, the forehead is lightly wrinkled, the eyebrows are raised a little or the mouth is closed. By and large, this face is dry, in that there is little to no perspiration visible. For as much as that is possible for a man who is always on edge, it is, you might say, the expression of Taylor's face at rest. The scroll plain tends to be the register employed in situations where Taylor isn't entirely sure yet what is happening around him. There is an openness to it, a lack of direction. The reason coach so often looks on edge, I think, is that he is always in a state of anticipation, is always ready to respond to something or someone: criticism by the press, job uncertainty, a comment by his wife, a demand from his daughter, trouble caused by his players. The scroll plain marks the moment where he still hasn't decided or hasn't had to decide how to respond, hasn't yet had a reason to assume he should be happy, or, rather more often, irritated. Indeed, if anything, it suggests a sense of possibility, whereas the other faces tend to indicate he has already made up his mind as to how he will feel, or what he can feel.

The floodplain, the plain formed through the sediment left by rivers flooding their banks, the terroir thickening with every course, for instance, the biting of the lower lip contracting the face, often marks the moment irritation gives way to feelings of frustration, resentment, melancholy and/or understanding; in any case, introspective sentiments. The outwash or glacier plain, where meltwater deposits ground rocks downhill, the stiffening of the lips dragging the whole face down to the chin, more often than not signals rising anger, indeed is generally the composition preceding one of Taylor's rarer outbursts – a heated argument with Tami, an intimidating lecture to a student, a grimace of pure agony to no one and anyone.

In the season one episode 'Ch-ch-ch-ch-changes' (episode 19), Taylor contemplates accepting a job offer coaching at university level in Austin. As he talks to his wife and his daughter early on in the decision-making process, his face looks like an open plain, long and still, his lips parted absentmindedly, like a passing stream, as if his thoughts are elsewhere, less with their immediate concerns than with his own aspirations. This expressive register is further accompanied by talk of the many opportunities Austin allegedly offers to his daughter, such as art and dance. But it is idle talk, ill prepared and inconclusive, repeating a single argument rather than developing it, using so many words for dance that he runs out of them, describing ballerinas as 'balleters', suggesting that he is trying to half-heartedly convince her but hasn't awoken to the repercussions himself. Later on in this episode, however, as his wife has expressed doubts and his daughter has shown her discontent, sulking like the teenager she is – eyes teary, lips pouted, screaming, blanking out her parents – Taylor's face takes on a slightly different form. In this particular instance the transition is made not within the shot but through editing, cutting from a close-up of his face with lips parted to a mid shot of him softly biting his lip. The extended effects of this small gesture are remarkable. In one cut, Taylor's face has lost much of its openness and length, is no longer a picture of contemplativeness. As he bites his lower lip he pulls his forehead, eyebrows and cheeks down while dragging his chin up, so that it points at us as opposed to downwards. You could say that this gesture contracts the entire mise-en-scène. For the moment he bites his lower lip he tilts his head downwards and slumps his shoulders, his body suddenly leaning towards the ground, the possibilities for the future literally limited, disappearing from view. So as to suggest the significance of his performance at this point, Taylor moves over towards the left of the image, his dark body contrasting with the bright light shining through the window behind him. Tellingly, his wife Tami remains standing towards the right of the frame, out of focus, her auburn blond hair and pastel shirt disappearing against the yellow curtains and cream walls. She looks at her husband, redirecting our gaze back at him (Figure 12.2).

It's a testament both to the series' attention to style and to Taylor's performance here that this smallest of gestures, the biting of a lip, intimates a turning point in the narrative as well as Taylor's state of mind. For this contraction of the face, the sedimentation of this landscape, sets in motion much of the season's remainder: a prolonged conflict between Taylor and his daughter Julie, introspection about his future, doubts about his priorities, worries, a renewed respect for his daughter as woman, and, finally, the decision to move to Austin on his own while his family stays behind in Dillon – for the time being. In one of the final scenes, Taylor and Julie find themselves in a car together on their way to a father–daughter dance at school. He is enthusiastic about the prospect of dancing with his daughter, though wary about her feelings. His angry daughter is intent on disliking it. As he parks the car, however, he initiates a discussion about their predicament: about

Figure 12.2   Taylor leans forward, slumping his shoulders and biting his lip and looks downwards as he turns inwards in disappointment with the world outside (*Friday Night Lights*, NBC, 2006–11)

his job, about her rights and plights as a 15-year-old child, the future of the family. His facial register during their conversation mirrors that of the episode's first half. Initially, his lips are pressed together, but lightly, not as casually as earlier, although certainly not as tersely as they are at other times – the third composition I will discuss shortly. As Julie tells him how she feels, however, as he is compelled to consider the thoughts of someone else, his expression changes: he begins ever so slightly to bite his lower lip. It is, on more than one level, an anticipation of what is to come: Taylor telling his daughter that he 'hears her' while smiling cautiously. Though he won't say it out loud until the season's final episode, a part of him has already decided what to do.

As seasons pass, Taylor's face sediments to the point that his expressive register narrows down considerably. It's not just that his expressions lose some of their intensity, sink deeper into the skin or conversely are swallowed up by skin, but also that the range dwindles: he shouts less, he smiles less. One of the reasons, visually, for this diminished expressiveness may well be the increasing extent to which Taylor is on edge, anticipates one or another emotional response to the world. As I mentioned above, the scroll plain, though it certainly doesn't disappear altogether, is in some scenarios displaced by either the flood plain or, still more, the outwash. It is as if the entire landscape is covered in sand and clay and grass and the original features are less and less visible. His face is a landscape weathered by heavy rains and heat waves, by flooding rivers and dissolving glaciers, one that has hardened to withstand them, that has gained in mass, in gravitas, what it may have lost in variety. Few instances in the series illustrate this weathering more clearly than Taylor's reaction to the possibility of his family picking up and leaving for Philadelphia. In contrast to the sense of inconsequentiality, of daydreaming, that Taylor's face displays with respect to the family's move to Austin, it betrays no such lingering thoughts the moment relocation to Philadelphia becomes an issue. The moment Tami tells him she has been offered a job there, in the episode tellingly titled 'Texas, Whatever' (season five, episode 12), his lips stiffen, gradually pressing down on one another, drawing his whole face to his chin. His face, though not sweaty, is certainly perspiring, a shimmer on his forehead and cheeks lit by a lamp light. You may, of course, say that his response differs because here it is not him but his wife who initiates the conversation, or that the job implies leaving beloved Texas, but as his friend Buddy Garrity asks him to consider another job in Dillon, one that by all accounts would be a promotion, his riposte is the same: his pushes his lips onto each other in agitation, his entire face sinking into his body, like a melting glacier sinking into earth. What this transition tells us, I would argue, is not that Taylor is a different person altogether, but that the experience of events like the first – and the many other we see him go through over the course of five seasons – sediments on his face. Chandler's performance suggests that he understands that no new turn opens up to a broad vista, widens the possibilities; it rather introduces other limitations (Figure 12.3).

I want to make one remark, finally, about the relationship between these three configurations, these three landscapes, and the micro movements that turn them into other sights. As I noted above, the scroll plain often turns into floodplain or outwash, but far less often into a scream, a cry or ecstatic laugh. It is as if the openness, the length, the relative looseness, does not translate well into grotesque contortions. It isn't transformed by seismic activity, by an earthquake, as manifestly as these terser compositions are. Indeed, I would be inclined to say that what makes these seismic shifts meaningful is the extent to which they appear convulsive, spasmodic: the extent to which they appear as visual shocks.

Figure 12.3 Taylor engages in a stare down with the future (*Friday Night Lights*, NBC, 2006–11)

Faces as landscapes

In the sections above I have tried to evaluate and contextualise Chandler's performance of coach Taylor, in particular the use of the mouth in relation to the rest of his face, by recourse to the metaphor of landscape, specifically the plain. Though my analyses here are far too concise to grasp the complexity of each gesture and the introduction of landscape theory has been little more than a glimpse of what the discussion may offer, I do hope that I have imparted to the reader both the extent of Chandler's achievement and a sense in which thinking about faces as landscapes, with a history and a spatial reality, affording some activities while rendering others unlikely or, if they do happen, extraordinary, may help us interpret screen performances, especially, I would suppose, televisual ones.

Bibliography

Affron, Charles (1977) *Star Acting: Gish, Garbo, Davis* (New York, NY: E.P. Dutton).

Aitkenhead, Decca (2009) 'Interview', *The Guardian*, 20 April.

Anon. (n.d.) 'Skippy', classicaustraliantv.com/Skippy.htm (accessed 8 March 2019).

Archive of American Television (2002) 'Interview with William Link', www.emmytvlegends.org/interviews/people/william-link (accessed 12 April 2017).

Auslander, Philip (1997) *From Acting to Performance: Essays in Modernism and Postmodernism* (New York: Routledge).

Austin-Smith, Brenda (2012) 'Acting Matters: Noting Performance in Three Films', in Aaron Taylor (ed.), *Theorizing Film Acting* (New York, NY: Routledge), pp. 19–32.

Bakhtin, Mikhail (1984) *Rabelais and His World* (Bloomington, IN: Indiana University Press).

Bandes, Susan (1996) 'Empathy, Narrative and Victim Impact Statements', *The University of Chicago Law Review*, 63.2, 361–412.

Bateson, Gregory (1955), 'A Theory of Play and Fantasy', reprinted in Gregory Bateson (1972), *Steps Towards an Ecology of Mind* (San Francisco, CA: Chandler), pp. 175–91.

Bateson, Gregory (1972) *Steps Towards an Ecology of Mind* (San Francisco, CA: Chandler).

BBC (2010) *Skippy: Australia's First Superstar* (tx., BBC4).

Bennett, James (2010) *Television Personalities: Stardom and the Small Screen* (Oxon: Routledge).

Berlant, Lauren (1997) *The Queen of America Goes to Washington City: Essays on Sex and Citizenship* (Durham: Duke University Press).

Berridge, Susan (2014) 'The Rachel: The Significance of Medium Specificity to Jennifer Aniston's Star Persona.' *https://cstonline.net/the-rachel-the-significance-of-medium-specificity-to-jennifer-anistons-star-persona-by-susan-berridge/* (accessed 8 March 2019).

Bignell, Jonathan (2005) 'And the Rest Is History: Lew Grade, Creation Narratives and Television Historiography', in C. Johnson and R. Turnock (eds), *ITV Cultures: Independent Television Over Fifty Years* (Buckingham: Open University Press), pp. 57–70.

Bignell, Jonathan (2011) '"Anything Can Happen in the Next Half-Hour": Gerry Anderson's Transnational Science Fiction', in T. Hochscherf and J. Leggott (eds), *British Science Fiction Film and Television: Critical Essays* (Jefferson: McFarland), pp. 73–84.

Bingham, Dennis (2010) 'Living Stories: Performance in the Contemporary Biopic', in Christine Cornea (ed.), *Genre and Performance: Film and Television* (Manchester: Manchester University Press), pp. 76–95.

Blanchet, Robert and Margrethe Bruun Vaage (2012) 'Don, Peggy, and Other Fictional Friends? Engaging with Characters in Television Series', *Projections*, 6.2, 18–41.

Bonner, Frances (2011) *Personality Presenters: Television's Intermediaries with Viewers* (Farnham: Ashgate).

Bouissac, Paul (1981) 'Behavior in Context: In What Sense Is a Circus Animal Performing?', in Thomas Sebeok and Robert Rosenthal (eds), *The Clever Hans Phenomenon: Communication with Horses, Whales, Apes and People* (New York, NY: New York Academy of Sciences), pp. 18–25.

Bould, Mark (2005) 'This Is the Modern World: The Prisoner, Authorship and Allegory', in Jonathan Bignell and Stephen Lacey (eds), *Popular Television Drama: Critical Perspectives* (Manchester: Manchester University Press), pp. 93–109.

Brand, Graham and Paddy Scannell (1991) 'Talk, Identity and Performance: The Tony Blackburn Show', in Paddy Scannell (ed.), *Broadcast Talk* (London: Sage), pp. 201–26.

Britton, Piers (2009) 'Design for Screen SF', in Mark Bould, Andrew Butler, Adam Roberts and Sherryl Vint (eds), *The Routledge Companion to Science Fiction* (London and New York: Routledge), pp. 341–49.

Brown, Tom (2012) *Breaking the Fourth Wall: Direct Address in the Cinema* (Edinburgh: Edinburgh University Press).

Bruun Vaage, Margrethe (2016) *The Antihero in American Television* (New York/Oxon: Routledge).

Butler, Judith (1990) *Gender Trouble* (London: Routledge).

Calfas, Jennifer (2013) 'This Is the End: Judd Apatow Taught Seth Rogen, Evan Goldberg the "Science" of Comedy', *Hollywood Reporter*, www.hollywoodreporter.com/news/is-end-judd-apatow-taught-562236 (accessed 13 June 2017).

Cantrell, Tom and Christopher Hogg (2016) 'Returning to an Old Question: What Do Television Actors Do When They Act?', *Critical Studies in Television*, 11.3, 283–98.

Cantrell, Tom and Christopher Hogg (2017) *Acting in British Television* (London: Palgrave Macmillan).

Cardwell, Sarah (2006) 'Television Aesthetics', *Critical Studies in Television*, 1.1, 72–80.

Carlson, Marvin (1996) *Performance: A Critical Introduction* (London: Routledge).

Carlson, Marvin (2003) *The Haunted Stage: The Theatre as Memory Machine* (Ann Arbor, MI: University of Michigan Press).

Carnicke, Sharon Marie (2010) 'Stanislavsky's System: Pathways for the Actor', in Alison Hodge (ed.), *Actor Training*, 2nd edn (London and New York: Routledge), pp. 99–116.

Cassidy, Gary and Simone Knox (2015) 'What Actors Do: Jennifer Aniston in Friends', *CST Online*, cstonline.net/what-actors-do-jennifer-aniston-in-friends-by-gary-cassidy-and-simone-knox (accessed 8 March 2019).

Caughie, John (2000) 'What Do Actors Do When They Act?', in Jonathan Bignell, Stephen Lacey and Madeleine MacMurraugh-Kavanagh (eds), *British Television Drama: Past, Present and Future* (London: Palgrave Macmillan), pp. 162–74.

Clayton, Alex (2012) 'Play-Acting: A Theory of Comedic Performance', in Aaron Taylor (ed.), *Theorizing Film Acting* (New York/Oxon: Routledge), pp. 47–61.

Comolli, Jean-Louis (1978) 'Historical Fiction: A Body Too Much', trans. B. Brewster, *Screen*, 19.2, 41–53 [originally published as (1977) 'Un corps en trop', *Cahiers du Cinema*, 278, 5–16].

Cooke, Lez (2015) *British Television Drama: A History*, 2nd edn (London: British Film Institute).

Cornea, Christine (2010) '2-D Performance and the Re-Animated Actor in Science Fiction Cinema', in C. Cornea (ed.), *Genre and Performance: Film and Television* (Manchester: Manchester University Press), pp. 148–65.

Corner, John (1999) *Critical Ideas in Television Studies* (Oxford: Oxford University Press).

Creeber, Glen (2004) *Serial Television: Big Drama on the Small Screen* (London: British Film Institute Publishing).

De Cordova, Richard (1991) 'Genre and Performance: An Overview', in Jeremy G. Butler (ed.), *Star Texts: Image and Performance in Film and Television* (Detroit, MI: Wayne State University Press), pp. 115–24.

Deleuze, Gilles (2005) *Cinema 1: The Movement Image* (London: Continuum).

Deleuze, Gilles and Felix Guattari ([1980] 2016) *A Thousand Plateaus: Capitalism and Schizophrenia* (London: Bloomsbury).

Dent, Suzie (2009) *What Made the Crocodile Cry?* (Oxford: Oxford University Press).

Derrida, Jacques (2002) 'The Animal that Therefore I Am (More to Follow)', *Critical Inquiry*, 28.2, 369–419.

Doane, Mary Ann (2003) 'The Close-Up: Scale and Detail in the Cinema', *Differences: A Journal of Feminist Cultural Studies*, 14.3, 89–111.

Dolan, Deirdre (2006) *Curb Your Enthusiasm: The Book* (London: Simon & Schuster).

Drake, Philip (2006) 'Reconceptualizing Screen Performance', *Journal of Film and Video*, 58.1/2, 84–94.

Drake, Philip (2016) 'Reframing Television Performance', *Journal of Film and Video*, 68.3–4, 6–17.

Duncan, Pansy (2011) 'Tears, Melodrama and "Heterosensibility" in *Letter to an Unknown Woman*', *Screen*, 52.2, 173–92.

Dyer, Richard (1974) *Frame Analysis* (New York, NY: Doubleday).

Dyer, Richard (1977) 'Entertainment and Utopia', *Movie*, 24, 2–13.

Dyer, Richard (1979) *Stars* (London: British Film Institute).

Dyer, Richard (1981) 'Introduction', in Richard Dyer, Christine Geraghty, Marion Jordan, Terry Lovell, Richard Paterson and John Stewart (eds), *Coronation Street* (London: British Film Institute), pp. 1–8.

Eco, Umberto (1977) 'Semiotics of Theatrical Performance', *The Drama Review*, 21, 107–17.

Elkins, James (2001) *Pictures & Tears* (London: Routledge).

Episodes (2017) coronationstreet.wikia.com/wiki/Category:Episodes (accessed 28 April 2017).

Equity (2015), 'Professionally Made Professionally Paid', www.equity.org.uk/campaigns/professionally-made-professionally-paid/ (accessed 22 May 2017).

Ewan, Vanessa and Debbie Green (2015) *Actor Movement: Expression of the Physical Being – A Movement Handbook for Actors* (London and New York: Bloomsbury).

Farber, Manny (1971) *Negative Space: Manny Farber on the Movies* (London: Studio Vista).

Forrest, David and Beth Johnson (2016a) 'Northern English Stardom', *Journal of Popular Television*, 4.2, 195–98.

Forrest, David and Beth Johnson (2016b) 'Lesley Sharp and the Alternative Geographies of Northern English Stardom', *Journal of Popular Television*, 4.2, 199–212.

Friedman, Sam, Dave O'Brien and Daniel Laurison (2016) '"Like Skydiving without a Parachute": How Class Origin Shapes Occupational Trajectories in British Acting', *Sociology*, 51.5, 1–19.

Geraghty, Christine (2010) 'Exhausted and Exhausting: Television Studies and British Soap Opera', *Critical Studies in Television*, 5.1, 82–96.

Gill, Rosalind and Andy Pratt (2008) 'Precarity and Cultural Work in the Social Factory? Immaterial Labour, Precariousness and Cultural Work', *Theory, Culture & Society*, 25.7–8, 1–30.

Goffman, Erving (1959) *The Presentation of Self in Everyday Life* (London, New York, Ontario and Auckland: Penguin).

Goffman, Erving (1969) *The Presentation of Self in Everyday Life* (London: Pelican Books).

Goffman, Erving (1974), *Frame Analysis* (New York: Doubleday).

Gordon, Robert (2006) *The Purpose of Playing: Modern Acting Theories in Perspective* (Ann Arbor, MI: The University of Michigan Press).

Gorton, Kristyn (2009) *Media Audiences: Television, Meaning and Emotion* (Edinburgh: Edinburgh University Press).

Grindstaff, Laura (2002) *The Money Shot: Trash, Class and the Making of TV Talk Shows* (Chicago: University of Chicago Press).

Haraway, Donna (1991) *Simians, Cyborgs, and Women: The Reinvention of Nature* (New York and London: Routledge).

Haraway, Donna (2003) *The Companion Species Manifesto: Dogs, People, and Significant Otherness* (Chicago: Prickly Paradigm).

Heath, Stephen (1981) *Questions of Cinema* (Basingstoke: Macmillan).

Hesmondhalgh, David and Sarah Baker (2008) 'Creative Work and Emotional Labour in the Television Industry', *Theory, Culture & Society*, 25.7–8, 97–118.

Hetzler, Eric T. (2007) 'Actors and Emotion in Performance', *Studies in Theatre and Performance*, 28.1, 59–78.

Hewett, Richard (2015) 'The Changing Determinants of UK Television Acting', *Critical Studies in Television*, 10.1, 73–90.

Hewett, Richard (2017) *The Changing Spaces of Television Acting* (Manchester: Manchester University Press).

Hill, Annette (2005a) 'Reality TV: Performance, Authenticity, and Television Audiences', in Jane Wasko (ed.), *A Companion to Television* (Malden, MA: Blackwell), pp. 449–67.

Hill, Annette (2005b) *Reality TV: Audiences and Popular Factual Entertainment* (London: Routledge).

Hilmes, Michele, et al. (2014) 'Rethinking Television: A Critical Symposium on the New Age of Episodic Narrative Storytelling,' Cineaste, 39.4, 26–38.

Hogan, Mike (2014) 'David Harewood, Homeland Actor, Says Nicholas Brody Deserves to Die', *Huffington Post*, 24 January, https://guce.huffpost.com/copyConsent?sessionId=3_cc-session_1b2018da-28f5-4270-b744-6b792fe013f7&inline=false&lang=en-us (accessed 26 September 2017).

Hoggart, Richard (1957) *The Uses of Literacy: Aspects of Working-Class Life with Special Reference to Publications and Entertainments* (London: Chatto & Windus).

Holmes, Su and Sean Redmond (2006) *Framing Celebrity: New Directions in Celebrity Culture* (Oxon: Routledge).

Holmes, Su and Sean Redmond (2007) *Stardom and Celebrity: A Reader* (London: Sage).

Jacobs, Jason (2001) 'Issues of Judgement and Value in Television Studies', *International Journal of Cultural Studies*, 4.4, 427–47.

Jacobs, Jason (2011) 'Inner and Outer in Homeland', *CST Online* blog, 2 December, http://cstonline.tv/inner-and-outer (accessed 26 September 2017).

Jacobs, Jason (2012) *Deadwood* (London: BFI/Palgrave Macmillan).

Jacobs, Jason and Steven Peacock (eds) (2013) *Television Aesthetics and Style* (New York and London: Bloomsbury).

Jaramillo, Deborah L. (2013) 'Rescuing Television from the "Cinematic": The Perils of Dismissing Television Style', in Jason Jacobs and Steven Peacock (eds), *Television Aesthetics and Style* (London: Bloomsbury), pp. 67–76.

Johnston, Sheila (2001) 'How to Make a Drama Out of a Crisis' *The Observer*, 18 November, p. 6.

Kavka, Misha (2008) *Reality Television, Affect and Intimacy* (Basingstoke: Palgrave).

Kemper, Tom (2010) *Hidden Talent: The Emergence of Hollywood Agents* (Berkeley and Los Angeles, CA: University of California Press).

Klevan, Andrew (2003) 'The Purpose of Plot and the Place of Joan Bennett in Fritz Lang's The Woman in the Window', *CineAction*, 62, 15–21.

Klevan, Andrew (2005) *Film Performance: From Achievement to Appreciation* (London: Wallflower Press).

Knox, Simone (2018) 'Exploring the Casting of British and Irish Actors in Contemporary US Film and Television' in Tom Cantrell and Christopher Hogg (eds), *Exploring Television Acting* (London: Bloomsbury), pp. 154–70.

Konjin, Elly (1995) 'Actors and Emotions: A Psychological Perspective', *Theatre Research International*, 20.2, 132–40.

Kotsko, Adam (2010) *Awkwardness: An Essay* (Ropley: John Hunt Publishing).

Kozloff, Sarah (1992) 'Narrative Theory and Television', in R.C. Allen (ed.), *Channels of Discourse, Reassembled: Television and Contemporary Criticism* (London: Routledge), pp. 67–100.

Kuleshov, Lev (1974) 'The Training of the Actor', in Ronald Levaco (trans. and ed.), *Kuleshov on Film* (Berkeley, CA: California University Press), pp. 99–115.

La Rivière, S. (2009) *Filmed in Supermarionation: A History of the Future* (Neshannock, PA: Hermes).

Langer, John (1981) 'Television's "Personality System,"', *Media, Culture and Society*, 3.4, 351–65.

Lawson, Mark (2013) 'Are We Really in a Second "Golden Age for Television"?', *The Guardian*, 23 May.

Lefebvre, Martin (2006) 'Introduction' in Martin Lefebvre (ed.), *Landscape and Film* (New York: Routledge), pp. xi–xxxi.

Leslie, Ian (2017) 'Watch It while It Lasts: Our Golden Age of Television', *Financial Times*, 13 April.

Logan, Elliot (2015) 'How Do We Write about Performance in Serial Television?', *Series*, 1.1, 27–38.

Logan, Elliot (2016) *Breaking Bad and Dignity* (Basingstoke: Palgrave Macmillan).

Lury, Karen (1995) 'Television Performance: Being, Acting and "Corpsing"', *New Formations*, 27, 114–27.

Lutz, Tom (1999) *Crying: The Natural & Cultural History of Tears* (New York: W.W. Norton & Co.).

Margolis, Ellen and Lissa Tyler Renaud (2010) 'Introduction', in Ellen Margolis and Lissa Tyler Renaud (eds), *The Politics of American Actor Training* (London: Routledge).

Marx, Karl (1954) *Capital Vol. 1* (London: Lawrence & Wishart).

McCabe, Janet and Kim Akass (eds) (2007) *Quality TV: Contemporary American Television and Beyond* (London: I.B. Tauris).

McLuhan, Marshall ([1964] 2009) *Understanding Media: The Extensions of Man* (London: Routledge).

Mills, Brett (2004) 'Comedy Verite: Contemporary Sitcom Form', *Screen*, 45, 63–78.

Mills, Brett (2005) *Television Sitcom* (London: British Film Institute).

Mills, Brett (2008) 'After the Interview', *Cinema Journal*, 47.2, 148–53.

Mills, Brett (2010) 'Invisible Television: The Programmes No-One Talks about Even Though Lots of People Watch Them', *Critical Studies in Television*, 5.1, 1–16.

Mills, Brett (2015) 'The Panel Show' in Glen Creeber (ed.), *The Television Genre Book* (Basingstoke: Palgrave Macmillan), pp. 109–10.

Mittell, Jason (2010) 'Previously On: Prime Time Serials and the Mechanics of Memory', in M. Grishakova and M. Ryan (eds), *Intermediality and Storytelling* (Berlin/New York: Walter de Gruyter GmbH & Co), pp. 78–98.

Mittell, Jason (2015) *Complex TV: The Poetics of Contemporary Television Storytelling* (New York: New York University Press).

Mittell, Jason (2017) 'The Ends of Serial Criticism', in Frank Kelleter (ed.), *Media of Serial Narrative* (Columbus, OH: The Ohio State University Press), pp. 169–82.

Moore, Paul (2004) 'Longing to Belong: Trained Actors' Attempts to Enter the Profession' (unpublished doctoral thesis, University of Sydney, Australia).

Nancy, Jean-Luc (2005) *The Ground of the Image* (New York: Fordham University Press).

Nannicelli, Ted (2016) *Appreciating the Art of Television: A Philosophical Perspective* (London/New York: Routledge).

Naremore, James (1988) *Acting in the Cinema* (Berkeley, CA: University of California Press).

Naremore, James (1990) *Acting in the Cinema* [paperback] (Berkeley, CA: University of California Press).

Neale, Steve and Frank Krutnik (1990) *Popular Film and Television Comedy* (London and New York: Routledge).

Nelson, Robin (2007) *State of Play: Contemporary 'High End' TV Drama* (Manchester: Manchester University Press).

Newcomb, Horace (1974) *TV: The Most Popular Art* (Garden City, NY: Anchor Books).

Newcomb, Horace and Paul Hirsch (1983) 'Television as a cultural forum', *Quarterly Review of Film Studies*, 8.3, 45–55.

Norton, Michael I., Jeanna H. Frost and Dan Ariely (2007) 'Less Is More: The Lure of Ambiguity, or Why Familiarity Breeds Contempt,' *Journal of Personality and Social Psychology*, 92.1, 97–105.

Nussbaum, Martha (2003) *Upheavals of Thought: The Intelligence of Emotions* (Cambridge: Cambridge University Press).

Peacock, Steven (2010) 'Borders and Boundaries in Deadwood', in Christine Cornea (ed.), *Genre and Performance: Film and Television* (Manchester: Manchester University Press), pp. 96–112.

Pearson, Roberta (2010) 'The Multiple Determinants of Television Acting', in Christine Cornea (ed.), *Genre and Performance: Film and Television* (Manchester: Manchester University Press), pp. 166–83.

Penrod, James (1974) *Movement for the Performing Artist* (Palo Alto, CA: National Press Books).

Pickering, Kenneth and Jayne Thompson (2013) *Naturalism in Theatre: Its Development and Legacy* (Hampshire: Palgrave Macmillan).

Piper, Helen (2004) 'Reality TV, Wife Swap and the Drama of Banality', *Screen*, 45.4, 273–86.

Pippin, Robert (2012) *Fatalism in American Film Noir: Some Cinematic Philosophy* (Charlottesville: University of Virginia Press).

Pixley, A. (1988) 'Everyman's Production', *Timescreen*, 11, 9–18.

Poniewozik, James (2016) '"This Is Us" Is Skillful, Shameless Tear-Jerking', *The New York Times*, 19 September.

Raney, Arthur A. (2004) 'Expanding Disposition Theory: Reconsidering Character Liking, Moral Evaluations, and Enjoyment,' *Communication Theory*, 14.4, 348–69.

Rawlins, Trevor (2012) 'Studying Acting: An Investigation into Contemporary Approaches to Professional Actor Training in the UK' (unpublished doctoral thesis, University of Reading, UK).

Richardson, M. (1991) 'F. A. B.', *Timescreen*, 6 (revised ed.), 4–7.

Robertson Wojcik, Pamela (2006) 'The Sound of Film Acting', *Journal of Film and Video*, 58.1/2, 71–83.

Rothman, William (2013) 'Justifying *Justified*', in Jason Jacobs and Steven Peacock (eds), *Television Aesthetics and Style* (London: Bloomsbury), pp. 175–184.

Scannell, Paddy (1996) *Radio, Television and Modern Life: A Phenomenological Approach* (Oxford: Blackwell).

Scannell, Paddy (2007) *Media and Communication* (London: Sage).

Scannell, Paddy (2013) *Television and the Meaning of Live: An Enquiry into the Human Situation* (Cambridge: Polity).

Sellers, Robert (2006) *Cult TV: The Golden Age of ITC* (London: Plexus).

Shacklock, Zoë (2016) 'Two of a Kind: Revaluing the Work of Acting Doubles in Orphan Black', *Journal of Film and Video*, 69, 69–82.

Shapiro, Stephen (2015) 'Homeland's Crisis of Middle-Class Transformation', *Cinema Journal*, 54.4, 152–58.

Skeggs, Beverly and Helen Wood (2012) *Reacting to Reality Television: Performance, Audience and Value* (London: Routledge).

Smith, Greg M. (2014) 'Coming Out of the Corner: The Challenges of a Broader Media Cognitivism', in Ted Nannicelli and Paul Taberham (eds), *Cognitive Media Theory* (New York/Oxon: Routledge), pp. 285–302.

Stage Castings, The (2015) 'How Actors Cope with Being Out of Work', *The Stage*, 12 February, http://castings.thestage.co.uk/audition-advice/blog/posts/how-actors-cope-with-being-out-of-work (accessed 22 May 2017).

Stanislavsky, Constantin ([1937] 1964) *An Actor Prepares*, trans. by Elizabeth Reynolds Hapgood (London: Geoffrey Bles).

Strasberg, Lee (1987) *A Dream of Passion: The Development of the Method* (New York: Plume).

Thompson, Ethan (2007) 'Comedy Verité? The Observational Documentary Meets the Televisual Sitcom', *Velvet Light Trap*, 60, 63–72.

Thompson, John O. (1978) 'Screen Acting and the Commutation Test', *Screen*, 19.2, 55–69.

Todorov, Tzvetan (1975) *The Fantastic: A Structural Approach to A Literary Genre*, trans. R. Howard (New York: Cornell University Press).

Tolson, Andrew (1991) 'Televised Chat and the Synthetic Personality' in Paddy Scannell (ed.), *Broadcast Talk* (London: Sage), pp. 178–200.

Trimble, Michael (2012) *Why Humans like to Cry: Tragedy, Evolution, and the Brain* (Oxford: Oxford University Press).

Tucker, Patrick (1994) *Secrets of Screen Acting* (New York/London: Routledge).

Ward, Amanda (1997) 'Dot: I'll Be Back!', *Daily Mirror*, 9 July.

Warhol, Robyn (2003) *Having a Good Cry: Effeminate Feelings and Pop-Culture Forms* (Columbus: Ohio State University Press).

Williams, Raymond (1974) *Television, Technology and Cultural Form* (London: Fontana).

Wood, Helen, Beverly Skeggs and Nancy Thumin (2009) '"It's Just Sad": Affect, Judgement, and Emotional Labour in "Reality" Television Viewing', in Stacy Gillis and Joanne Hollows (eds), *Feminism, Domesticity, and Popular Culture* (London: Routledge), pp. 135–50.

Wright Wexman, Virginia (1980) 'The Rhetoric of Cinematic Improvisation', *Cinema Journal*, 20.1, 29–41.

Zborowski, James (2016a) 'Soap Stats' Between Sympathy and Detachment, 6 December, https://betweensympathyanddetachment.wordpress.com/2016/12/06/soap-stats/ (accessed 28 April 2017).

Zborowski, James (2016b) 'Representing the Everyday in Coronation Street (1960 and 2013)' *British Television Drama*, www.britishtelevisiondrama.org.uk/?p=5922 (accessed 26 April 2017).

Index